Information Technology and Law Series

Volume 37

Editor-in-Chief

Simone van der Hof, eLaw (Center for Law and Digital Technologies), Leiden University, Leiden, The Netherlands

Series Editors

Jef Ausloos, University of Amsterdam, Amsterdam, The Netherlands

Stephan Dreyer, Leibniz-Institut für Medienforschung, Hamburg, Germany

Gloria González Fuster, Law, Science, Technology & Society Studies (LSTS), Vrije Universiteit Brussel (VUB), Brussels, Belgium

Inge Graef, University of Tilburg, Tilburg, The Netherlands

Aleksandra Kuczerawy, Centre for IT and IP, KU Leuven, Leuven, Belgium

Eva Lievens, Faculty of Law, Law & Technology, Ghent University, Ghent, Belgium

Aurelia Tamo-Larrieux, Faculty of Law, Maastricht University, Maastricht, The Netherlands

The *Information Technology & Law Series* was an initiative of IT *e* R, the national programme for information Technology and Law, which was a research programme set up by the Dutch government and The Netherlands Organisation for Scientific Research (NWO) in The Hague. Since 1995 IT *e* R has published all of its research results in its own book series. In 2002 IT *e* R launched the present internationally orientated and English language *Information Technology & Law Series*. This well-established series deals with the implications of information technology for legal systems and institutions. Manuscripts and related correspondence can be sent to the Series' Editorial Office, which will also gladly provide more information concerning editorial standards and procedures.

Francisco Pereira Coutinho · Martinho Lucas Pires ·
Bernardo Correia Barradas
Editors

Blockchain and the Law

Dogmatics and Dynamics

Editors
Francisco Pereira Coutinho
Department of Law
Nova School of Law
Lisbon, Portugal

Martinho Lucas Pires
Department of Law
Universidade Católica Portuguesa
Lisbon, Portugal

Bernardo Correia Barradas
World Bank Group
Lisbon, Portugal

ISSN 1570-2782   ISSN 2215-1966 (electronic)
Information Technology and Law Series
ISBN 978-94-6265-578-2   ISBN 978-94-6265-579-9 (eBook)
https://doi.org/10.1007/978-94-6265-579-9

Published by T.M.C. ASSER PRESS, The Hague, The Netherlands www.asserpress.nl
Produced and distributed for T.M.C. ASSER PRESS by Springer-Verlag Berlin Heidelberg

© T.M.C. ASSER PRESS and the author(s) 2024

No part of this work may be reproduced, stored in a retrieval system, or transmitted in any form or by any means, electronic, mechanical, photocopying, microfilming, recording or otherwise, without written permission from the Publisher, with the exception of any material supplied specifically for the purpose of being entered and executed on a computer system, for exclusive use by the purchaser of the work.
The use of general descriptive names, registered names, trademarks, service marks, etc. in this publication does not imply, even in the absence of a specific statement, that such names are exempt from the relevant protective laws and regulations and therefore free for general use.

This T.M.C. ASSER PRESS imprint is published by the registered company Springer-Verlag GmbH, DE, part of Springer Nature.
The registered company address is: Heidelberger Platz 3, 14197 Berlin, Germany

Paper in this product is recyclable.

Preface

Blockchains and distributed ledger technologies (DLTs) are one of the major technological developments of our times. DLTs are computer protocols designed for storing and managing data in a distributed manner, through different participants connected in a peer-to-peer, decentralized digital structure. Blockchain is the name attributed to a type of DLT that were inspired by the protocol behind the Bitcoin cryptocurrency, the first "trustless-based" payment system in the world, as per the words of Satoshi Nakamoto, its founder. The monetary value and volatility of the cryptocurrency were first to capture the attention of public opinion; since then, new technologies improving and expanding the capacity of the "genesis" protocol have appeared, morphing blockchain and crypto into a technological sector called "blockchain".

Currently, there is an enormous offer (and continuously growing one) of projects in the blockchain landscape. While new sectors have emerged with DeFi protocols and products fighting to establish a solid and profitable decentralized financial market, crypto exchanges have become regulated and incorporated in various jurisdictions. Crypto is being used as an alternative savings and investment product, while so-called traditional "legacy" institutions—e.g., banks, investment funds, and public administrations—are developing or adopting applications to enter the market and offer DLT-based services. Long gone are the days in which blockchain and crypto were looked on as dangerous products, with the sole purpose of escaping law enforcement, and disguised as simplistic claims of libertarian desires and cyber-punkish worlds. Crypto is now mainstream.

Proponents of the technology argue that blockchain has the potential to substantially alter commercial relations, either those that are business-to-business or those that are business-to-consumer, by switching commercial trust from market agents and intermediaries to computer programs based on algorithmic trust that run on protocols distributed by computers all over the world. Blockchain technology has brought about a paradigm shift in the way we think about and exchange value. Its potential for innovation and disruption has been widely recognized across various industries. However, any new technology brings legal and regulatory challenges that must be navigated.

The law and blockchain have a difficult relationship: the core principles of privacy, decentralization, and automation of the technology conflict with classic juridical standards concerning stability and liability of legal relations, although new laws have been enacted during the past years to help.

From 2018 to 2020, the three of us have organized at Nova Law School three short post-graduate courses on the legal impacts of blockchain technology. These courses were the first of their kind to be offered in Portuguese universities, and a variety of Portuguese and international academics, entrepreneurs, lawyers, and regulators participated. This book condenses a "best of" of the sessions taught in those three courses, dealing with issues relating to the convoluted application of legal standards and rules to blockchain technology.

Staying true to the international structure of the courses, the book contains eight chapters dedicated to general and more specific intersections between blockchain or DLT-based innovations and the law.

The book starts with Martinho Lucas Pires' chapter entitled "Blockchain and the Law: Setting the Floor" (Chap. 1). Call it a general introduction to the state of the art in the blockchain and law debate and an attempt to summarize the real challenges and the not so real hypes concerning the relationship between blockchain and its innovations and current legal standards.

Nelson M. Rosario in Chap. 2 adventures into assessing the promises and shortfalls of blockchain-based structures for voting. In Chap. 3, Renato Gomide Martinez de Almeida discusses the use of blockchain in transitional justice processes, by exploring its application in criminal prosecutions, truth commissions, reparations, in restoring democratic institutions and in reconciliation. In Chap. 4, Marta Carmo looks at the evolution of the tax treatment of crypto assets in Portugal, explaining and deconstructing the approaches by the Portuguese tax authorities that lead to the country in certain media circles being named as a "crypto tax haven".

In Chap. 5, George Daniel Raath dives into the topic of blockchain in the execution of judgments in the context of South Africa.

João Vieira dos Santos and Guilherme Maia in Chap. 6 consider the legislative initiative of the European Union in setting up a DLT Pilot Regime and assess its possible impact on DeFi products. Anjeza Beja and Bernardo Correia Barradas in Chap. 7 look at the novel central bank digital currency, or CBDC, and its implications on anonymity. And, in Chap. 8, Ágata Ferreira concludes the book by discussing the specific relationship between smart contracts and the law, looking at the advantages of smart contracts and addressing the scholarly legal debate surrounding their implementation and use.

All in all, this book provides a comprehensive overview of the intersection between blockchain and the law. It delves into the legal implications of blockchain technology and how it is being treated under existing legal frameworks. It also explores the potential of blockchain to reshape the legal landscape and how the law will continue to evolve in response to the growing use of blockchain. The book offers valuable insights into the complex relationship between blockchain and the law. It provides a well-rounded understanding of the legal implications of blockchain and how it can be leveraged for the benefit of society.

We would like to acknowledge and thank all the authors for their outstanding contribution to this volume and especially thank Frank Bakker and Kiki van Gurp from T.M.C. Asser Press for accepting this volume for publication and all the patience with the editing process. We wish all our readers an enjoyable and rewarding reading.

Lisbon, Portugal
May 2023

Francisco Pereira Coutinho
Bernardo Correia Barradas
Martinho Lucas Pires

Contents

1 **Blockchain and the Law: Setting the Floor** 1
  Martinho Lucas Pires

2 **Blockchain and Elections: Opportunity and Peril** 17
  Nelson Rosario

3 **What Are the Limits of Blockchain? Considerations on the Use of Blockchain in Transitional Justice Processes** 29
  Renato Gomide M. de Almeida

4 **Taxing Crypto-Assets—The Portuguese Perspective** 51
  Marta Carmo

5 **Blockchain Execution of Judgements—A Possibility in South Africa?** 75
  George Daniel Raath

6 **The DLT Pilot Regime and DeFi** 89
  Guilherme Maia and João Vieira dos Santos

7 **Central Bank Digital Currency: A Focus on Anonymity** 107
  Anjeza Beja and Bernardo Correia Barradas

8 **Smart Contracts and the Law** 125
  Agata Ferreira

Editors and Contributors

About the Editors

Francisco Pereira Coutinho is Associate Professor at NOVA School of Law in Lisbon, Portugal, where he directs the Observatories of Portuguese Legislation and Data Protection Law, as well as the Moot Court Program. Specialized in Public International Law and European Constitutional Law, his research and teaching interests further encompass comparative constitutional law, data protection law, and media law.

Martinho Lucas Pires is Teaching Assistant in the Law School of Universidade Católica Portuguesa, in Lisbon, Portugal, and of Counsel at DLA Piper. His research interests include financial regulation, regulation of crypto and DLTs, Fintech, and constitutional theory.

Bernardo Correia Barradas is Lawyer and Consultant in the Payment Systems Development Group of the World Bank Group, currently based in Aix-en-Provence, France. His research interests include new technologies applied to payments, CBDC, financial inclusion, and financial market infrastructures.

Contributors

Anjeza Beja Prokop Myzegari, Tirana, Albania

Marta Carmo Banco BPI, Lisbon, Portugal

Bernardo Correia Barradas Aix-en-Provence, France

Agata Ferreira Department of Administrative Law and Public Policy, Warsaw University of Technology, Faculty of Administration and Social Sciences, Warsaw, Poland

Renato Gomide M. de Almeida São Paulo, SP, Brazil

Martinho Lucas Pires Católica Law School, Lisbon, Portugal

Guilherme Maia BCAS, Msida, MSD 1825, Msida, Malta

George Daniel Raath MinterEllison, Canberra, Australia

Nelson Rosario Rosario Tech Law, LLC; Illinois Tech, Chicago-Kent College of Law, Chicago, IL, USA

João Vieira dos Santos Lusófona University and Portuguese Securities Market Commission (CMVM), Lisbon, Portugal

Abbreviations

AI	Artificial intelligence
AML/CFT	Anti-money laundering/countering the financing of terrorism
AMM	Automated market maker
API	Application programming interfaces
ATM	Automated teller machine
BIS	Bank for International Settlements
BoE	Bank of England
BTA	(Illinois) Blockchain Technology Act
CARF	Crypto-asset reporting framework
CBDC	Central bank digital currency
CBOB	Central Bank of the Bahamas
CBU	Central Bank of Uruguay
CCP	Central counterparty
CDD	Customer due diligence
CFTC	(US) Commodity Futures Trading Commission
CIT	Corporate income tax
CJEU	Court of Justice of the European Union (see ECJ)
CMVM	(Portuguese) Securities Market Commission
CPU	Central processing unit
CSD	Central securities depository
DAO	Decentralized autonomous organizations
dapps	Decentralized applications
DeFI	Decentralized finance
DEX	Exchange
DLT	Distributed ledger technology
EBA	European Banking Authority
ECB	European Central Bank
ECCB	Eastern Caribbean Central Bank
ECCU	Eastern Caribbean Currency Union
ECJ	European Court of Justice (see CJEU)
EEA	European Economic Area

ESCB	European System of Central Banks
ESMA	European Securities and Markets Authority
EU	European Union
FATF	Financial Action Task Force
FED	Federal Reserve System (USA)
FSB	Financial Stability Board
GA	General Assembly (UN)
GPT	General-purpose technology
GTL	(Portuguese) General Tax Law
IAS	International Accounting Standard
IBM	International Business Machines (Corporation)
ICC	International Criminal Court
ICO	Initial Coin Offering
ICTJ	International Center for Transitional Justice
ID	Identification
IFRIC	International Financial Reporting Interpretations Committee
IFRS	International Financial Reporting Standards
IMF	International Monetary Fund
IMT	Municipal Tax on Onerous Transfer of Real Estate (Portugal)
International IDEA	International Institute for Democracy and Electoral Assistance
IoT	Internet of Things
KYC	Know your customer
MiCA	Market in Crypto-Assets (Regulation)
MiFID II	.
ML/FT	Money laundering/financing of terrorism
MTF	Multilateral trading facilities
NFT	Non-fungible token
NGO	Non-governmental organization
OCC	(Portuguese) Chartered Accountants Association
OCHA	United Nations Office for the Coordination of Humanitarian Affairs
OECD	Organization for Economic Cooperation and Development
OTF	Organized trading facilities
P2P	People-to-people
PBC	People's Bank of China
PIT	Personal income tax
PoW	Proof-of-work
SR	Sveriges Riksbank (Sweden)
SS	Settlement system
TA	(Portuguese) Tax authority
TC	Truth commissions
TFV	Trust fund
TJ	Transitional justice
TRC	Truth and reconciliation commissions

Abbreviations

TSS	Trading and settlement system
UK	United Kingdom of Great Britain and Northern Ireland
UKJT	UK Jurisdiction Taskforce
UN	United Nations
UNHCR	Office of the High Commissioner for Refugees (United Nations)
US	United States (of America)
USA	United States of America
USD	United States Dollar
UTP	Unofficial Truth Projects
VAT	Value-added tax
WFP	World Food Program

Chapter 1
Blockchain and the Law: Setting the Floor

Martinho Lucas Pires

Contents

1.1	Introduction	2
1.2	The Legal "Phenomenon"	4
1.3	The Blockchain "Phenomenon"	6
1.4	The Law in Blockchain	9
1.5	The Law for Blockchain	11
1.6	Conclusions	12
References		13

Abstract Blockchain and the law has over the past years become a field in which a multitude of developments have taken place, and the tendency is for that progress to continue. However, due to the seemingly never-ending stream of publications and ongoing developments in the blockchain and crypto markets, the essence of the topic—concerning the legal effects and consequences of the growing adoption of blockchain based-technologies and their applications such as crypto and smart contracts—risks becoming a little lost. This chapter discusses the problem of blockchain and the law at its most basic and fundamental levels, and in so doing considers the essential traits of blockchain and the essential traits of law, looking into how both are related, and how new developments should be deployed, in a way that is faithful to the technology's promise and capacity, and to the law's function as the main tool for social order.

Keywords Blockchain · Law · Regulation · Crypto · Smart contracts

---

M. Lucas Pires (✉)
Católica Law School, Palma de Cima, 1649-023 Lisbon, Portugal
e-mail: martinholucaspires@ucp.pt

1.1 Introduction

In the first quarter of 2021, Bitcoin, Ethereum, and almost every other crypto asset posted new all-time prices in what was seen as the beginning of a new bull market. Enthusiasts lauded the price rise and attributed it to the growing adoption of crypto and blockchain protocols by corporations and investors alike from the whole economic spectrum: corporations like Tesla and Micro Strategy, traditional legacy players such as banks, and even States such as Ethiopia. Roughly four years after the last bull market and the end of the "hype cycle" of blockchain, 2021 seemed poised to be the year when the technology would start to succeed. According to proponents, its success would herald a new era of digitalization of transfers of value.[1]

Ironically, another crypto-related phenomenon peaked during those difficult years of the bear market (roughly between the beginning of 2018 and the end of 2020, if one is to draw the correct historical lines). The rise of the crypto-economy led to a lively and intense strand of publishing and research on all things crypto and blockchain. Legal academia was no exception. There were countless articles and books, thesis (master's and Ph.D. level), reports, and working papers published on the matter.[2] Large and significantly participated blockchain conferences were organized, and new specialized journals were inaugurated.[3] Observatories[4] and working groups were established, research projects implemented, associations built, and courses opened at major universities. In just about three years, the topic of blockchain and the law surfed the wave of the phenomenon and had a significant boom, becoming an autonomous strand of legal teaching and research.

It is typical for lawyers and law researchers to hop in the wagon of technological developments whenever something seems sufficiently innovative to bring profound changes to how people interact. However, no one could have predicted the quick and intense growth of the area of the studies of blockchain and the law. Legal scholars received blockchain enthusiastically, launching themselves in adopting and researching this new technology and new form of economy, organization, and community-building. Blockchain and distributed ledger technologies seemed to herald a new era of social disruption based on solid and provocative concepts such as decentralization, smart contracts, and digital money. Despite adoption being slow, the promise of blockchain to digitalize value and its allure was so appellative that two of the most prominent authors of the movement decided to coin a term to describe the new form of law that would deal with the developments of blockchain technology – developments that, according to said authors, "traditional" law could not answer. "Lex cryptographica" was to be the new, "revolutionary"

---

[1] Kapasi 2021; Spilka 2021.

[2] The main books would be De Fillipi and Wright 2018; Fink 2018; Magnusson 2020; Hacker et al 2019 and Fox and Green 2019.

[3] Stanford Journal of Blockchain Law and Policy, https://stanford-jblp.pubpub.org/, accessed 31 October 2022.

[4] The EU Blockchain Observatory and Forum, https://www.eublockchainforum.eu/, accessed 31 October 2022.

law that would deal with automated contracts (whatever those were), autonomous organizations (idem), and the emergent token economy.[5]

Being part of the "movement" of blockchain and the law by researching, counselling, organizing courses, teaching courses, participating in conferences, and publishing on the topic, I must confess, almost four years since the start, that the enthusiasm about blockchain and the law always looked, in a way, a bit overhyped. Call me a realist (I accept it gladly) or a conservative (less so). However, it always seemed that arguments supporting the exceptionality of blockchain as a social phenomenon and its "revolutionary" legal character went beyond reason and more into the realm of philosophical speculation. It is true that the blockchain phenomenon, despite its complexities and contradictions, poses challenges to the operation of law.[6] It is also true that traditional legal solutions may not be the best way (nor, in fact, a *possible* way) to deal with blockchain and its issues.

Nevertheless, the law is also a rich and complex reality, composed of rules and their implementation, values, and social perception, of its politics. The terms and principles governing legal relations, their method, and dogmatic are as old as humanity's existence. The law is, in this regard, the greatest legacy institution there is. Legal terms and principles have evolved, accompanying, and accommodating our technical progress. Law is a phenomenon of social order and power that is impossible to avoid or overcome. That is why I argue that the legal challenges of blockchain technology are not completely "revolutionary" nor "innovative" per se, from a legal, dogmatic perspective, and most of them can be resumed and solved in terms of the frameworks of contract and liability. The other cases must be tackled through innovative solutions that respect the purpose and structure of the technology and the good developments that it brings. Another question concerns the capacity of State-made law and national authorities to enforce this market's regulation effectively. However, that is another issue, one not of law but of administrative action and adjudication.

The present introductory chapter develops this argument, in an attempt to, in 2022, when the boom almost went bust, frame the basic structural features of the connection between the phenomena of blockchain and the law, the tensions they raise, and to assess them from a critical point of view. Call it an attempt to draw a framework for understanding the basics of what is at stake within the problem of blockchain and the law.

For this purpose, I divide the contribution into four sections. The first section concerns a discussion of law and its phenomenon, explaining what law is essentially about and how is represented and expressed from a practical and realist perspective. The second section discusses the blockchain phenomenon, its many dimensions, contradictions, and innovations. The third section looks at the supposedly disruptive effect of innovations brought by DLTs and assesses them from the more structural perspective of law—in other words, it looks at the law existent in the blockchain phenomenon. The fourth section considers what legal changes the blockchain market

---

[5] De Fillipi and Wright 2018, 5–9.

[6] See, for example, EU Blockchain Observatory and Forum 2020a.

could and should accommodate; changes that would bring more conceptual and practical precision.

In the end, I hope to be able to explain and expose the significant legal challenges posed by the growing adoption of blockchain and present some thoughts on how law and the market can evolve from here. Given what has already been written on the topic, I will be concise and generalize some points, particularly when discussing crypto assets and smart contracts—other chapters in this book will take on those subjects in greater depth.

1.2 The Legal "Phenomenon"

If we wish to be simple—and, given the limits of our chapter, simplicity is a must—an immediate definition of what the law is would be more or less the following. Law is a system of rules in force within a particular community, laid down by an authoritative institution, which organizes individual and collective behavior according to a criterion of justice. Of course, it is difficult for any definition of the law, even the most straightforward and immediate, not to succumb to the second reading when we assess all that the definition includes. What do we mean by rules? What is a community? What is an authoritative institution? Why do we need to organize our behavior? Moreover, what is justice? Let us then deconstruct the sentence for a more comprehensive view (but still somewhat simple and practical) of the legal phenomenon.

Law is a system of rules—by this expression, I mean that law is a set of norms of linguistic statements that, in general, are prescriptions.[7] They invite people to do or not to do as expressed in the statement and declare what sanctions should be applied to those who do not comply. Law is, therefore, concerned with what people ought to do and, conversely, with what they ought not to do.[8] Norms can also represent axiological declarations or mere enunciations. For the sake of simplicity, let us acknowledge and consider the prescriptive face of norms.

The purpose of legal norms is to establish some order, that is, some space of existence where expectations concerning external individual and collective behavior are set.[9] Order and predictability allow for better security, for people get to know better what to expect from one another when interacting and can, therefore, define their existential actions in accordance.

Order, stated as such, is a very formal and empty concept. Any legal order (the set of rules of a given political community or community) is established under specific values and principles.[10] Values and principles are choices, made by the rulers—those who enact the law—concerning the quality of the order that the community ought to have. Values and principles are choices made by the rulers—those who enact the

---

[7] Hughes 1968, pp. 411–439.
[8] Kelsen 1967, pp. 33–35.
[9] Raz 1979, p. 51, and Luhmann 2004, p. 143.
[10] Allot 1999, pp. 31–32.

law—concerning the quality of the order the community ought to have. Justice—that at a very functional and open level can be defined as the duty to provide to each its own due[11]—is the orientation that guides the choices of rulers, the ultimate quality criteria. Justice is a strong concept, and a discussion of its core goes beyond this scope; suffice to say that it usually represents an idea of what is "right" (e.g., correct).[12] Law sets and represents (or, at least, aspires to set and represent) the *rightest* form of social order.

The following considerations are the most complicated, for they deal with the question of law being the product of a will. When stating that an authoritative institution enacts a law, what I mean (and what, I suspect, readers immediately assume) is that law is a form of political action.[13] That is, the law is made by an institution that has the power to rule over society, a power that is not only genuine but legitimate, that is, justified in light of social criteria—for example, because of it is expressed by a democratically-elected body. It is also part of the ruling power of the institutions responsible for upholding the law by making its results achievable and applying sanctions to those who fail to comply with the law.

There is a more social view of the law that is not so much concerned with exercising the power of top-down ruling authority but with the collective expression of law emerging from the community through routine interactions. The social expression of law is called custom, whereby people act according to a collective perception of what is legally mandatory, which can be at odds with positive law (e.g., law made by the rulers).[14] Legal orders, and their rulers, not only admit the existence of such legal sources but also acknowledge individual autonomy in the definition of terms of social interactions by the mode of private law. Private law is (or can be generally described as) the rules dealing with relations between individuals outside of a relation of political (ruling) authority.[15] The liberal premise of private law is that people can do whatever they want—i.e., they can set the terms of their legal relations as they wish—within limits set out by the law. There is a space where the community—its politics—does not prescribe a specific outcome nor set a preferred action method. Hence a significant role is given to individual consent in the realm of private law under certain principles. This includes recognizing forms of collective action (e.g., associations, enterprises, corporations) and contractual freedom to set the terms of interaction, but also specific forms of resolving litigation (arbitration, for example).

Therefore, we can say that the legal phenomenon consists of rules of conduct, their meaning, application, and upholding within a political community. Some rules are politically made; they are enacted, and the ruling authorities of the community guarantee their efficacy. Other rules, however, can be born out of a collective agreement or through repeated and regular interaction with the community members. A

---

[11] Aristotle 2002, Book V.
[12] Ibid.
[13] Alexander 2019, p. 355, and Cerar 2009, pp. 21–22.
[14] Mangabeira Unger 1976, pp. 49–50 and 127.
[15] Kelsen 1967, p. 280.

(liberal) legal order will have imperative ruling while allowing for large spaces of private ordering.

The legal operation (enactment, enforcement, and adjudication of the law) is done by each community within its territory (it is also possible for communities to allow other external entities to exercise legal powers over them, e.g., federations, international organizations under certain conditions and concerning limited policy areas). The set of fundamental legal problems and general answers are fundamentally the same in every State: relations between the collectivity and the political power are ruled by a constitution that establishes the rights and duties of the governed and the government. In contrast, relations between the community members follow the logic of free commitment under knowledgeable consent, with liability rules applicable for actions that question or violate commitments people entered with one another. There are many differences among each community—for example: what constitutes free commitment under knowledgeable consent, under which condition liability arises, and the degree of fault that is necessary to show, among other rules—because each community shall have its version of what is right and just, within its specific context. From a realist and a rather essentialist point of view, these are the primary structural points concerning law to fulfil its function of orienting behavior and enabling an orderly life.

1.3 The Blockchain "Phenomenon"

Blockchain is, first and foremost, a technology: a form "of practical, especially industrial, use of scientific discoveries"[16] for human development. It is a digital technology based upon computer technology and created for communicating information or data: "a multi-party system in which participants reach agreement over a set of shared data and its validity, in the absence of a central coordinator".[17] Blockchain is a technology for organizing and acting on digital data in a distributed system among different computers.

The most famous blockchain protocol is Bitcoin. The technical and ideological importance of the Bitcoin protocol should not be underestimated. Most of what is described as blockchain, its principles, benefits, and its problems and risks refer to the Bitcoin protocol's characteristics. The basic assumption is widely known: Bitcoin established a system for exchanging value (in its case, value refers to information regarding the property of a token known as Bitcoin that would serve as a global digital currency) among people through a decentralized peer-to-peer network. By decentralized, the idea is that there is no intermediary between the parties to the transaction: the network and its users control the process. Two people in the network with cryptographic identities agree to exchange their tokens and communicate their intention

---

[16] Cambridge Dictionary, 'Technology' https://dictionary.cambridge.org/us/dictionary/english/technology, accessed 31 October 2022.
[17] Rauchs et al. 2018, p. 22.

to the network by signing off on the transaction with their respective cryptographic identities. The network receives all transactions, and as soon as another user (called the miner) solves an automatically generated puzzle, a new block in the network is formed, and the transaction is inserted, timestamped, and put in chronological order. As soon as users confirm the block within the chain, the transaction is sealed, and anyone can see the block and check the information but not change it.[18]

The decentralization claim in Bitcoin has a second, significant meaning: the lack of ruling authority, such as a board of directors. The anonymity of its creator, Satoshi Nakamoto, helped to insufflate Bitcoin and to leave its management in the hands of the community. This means that protocol governance is highly frictional, as some events have shown, such as the SegWit update and the "blocksize wars",[19] and it can lead to hard forks, a separation of the protocol in two, where one part runs with the same rules, and other remains unchanged. Blockchain then came to signify a distributed technology with decentralization features, both technically and in terms of governance.

The majority of blockchain protocols that followed were developed differently, even if they took technical and ideological inspiration from Bitcoin and Nakamoto. A common framework of terms is used to distinguish between first-generation protocols (basically, Bitcoin) from second-generation (Ethereum) and third-generation (e.g., Cardano, Algorand).[20] The second generation of protocols yielded the era of "smart contracts," that is, programmable digital scripts that enable the creation of applications such as programs, protocols, and tokens. The third generation brought staking: a change to the consensus mechanism, whereby consensus does not depend on computational energy (as in Bitcoin, the proof of work consensus mechanism) but rather on the number of tokens that participants in the network have. There are also other protocols with access to which are not public (anyone can download the protocol and run it, as long as it has sufficient computing power to do so) but private (people need permission from the protocol's creators to join the network).[21]

For second and third-generation protocols, decentralization exists only in the first meaning. In the second meaning, its existence is less obvious. Although there is, supposedly, no central formal authority over the protocols, the protocol founders are "public" figures, usually with a solid online presence and important commercial aspirations. In some cases, protocols are supported by private entities (foundations, companies) with sufficient funds and human capital (mainly developers) to act and influence how the protocol is to be run. For example, the way the Ethereum foundation supported a hard fork to change a bug in a code[22] or the way Justin Sun acquired Steem protocol are good examples of this influence.[23] Therefore, it is unclear that they are as decentralized as aspired from a ruling perspective.

---

[18] For a description of the functioning of Bitcoin, see Narayanan et al 2016, pp. 1–54.
[19] Bier 2021.
[20] See, in general, Ors 2021.
[21] EU Blockchain Observatory and Forum 2020b, 7–16.
[22] Meier and Schluppi 2019, 27–44.
[23] Copeland 2020.

So far, I have not touched on tokens and cryptocurrencies. The Bitcoin protocol generated, as an incentive, its token: a digital representation of value stored in a user's wallet and registered in the protocol. The token's purpose was to be used as a payment mechanism, as a currency; hence, all protocol tokens (stored through cryptographic means) were called cryptocurrencies. In the crypto economy, the token serves two functions: an economic incentive for the decentralized consensus mechanism to work and a product—in other words, as the element that makes the protocol valuable.

The value of Bitcoin comes from its use as a currency—payment mechanism, story of value or unit of account (the jury is still out there for which one is Bitcoin better equipped to handle). However, other tokens have different uses. For example, Ether is used to pay for storage pace and transaction fees in the Ethereum network; Filecoin enables users to store data in the Filecoin protocol; Tezos enables holders to vote on the protocol's governance; Ada, when staked with a pool, generates dividends for its holders. Some tokens represent securities that are used as collateral that are issued through public offerings. In short: tokens represent different value types, such as rights, currencies, agreements, property… To consider them solely as currencies is inadequate, thereby the preferred term being crypto assets, given the multiplicity of utilities they can have.

Nevertheless, market practice divided tokens in terms of function. This division has subsequently been adopted by public institutions, including EU regulators, in enacting the proposed Market in Crypto Assets Regulation (MiCA).[24] The classic functional division is between currency tokens (tokens resembling currencies), security tokens (tokens resembling securities), and utility tokens (basically, the rest). The classification is somewhat simplistic and too artificial, for there are tokens that can easily fit more than one category. For example, a token created to function as a currency, but providing its users with dividends when staked, is really (or should be considered) just a currency token. Is a utility token issued through a public offering just a utility token? These categories worked for illustration purposes when the market was beginning to appear in 2017; in 2021, with some protocols working in full force and more users going for decentralized finance (DeFi) protocols and tokens, it seems complicated to sustain such classification.[25]

Tokens are, arguably, the most known facet of the blockchain phenomenon. They are the central pieces for the functioning of the protocols and have become an easy source of liquidity. Tokens can be exchanged in seconds, between people, with minimum or no intervention from intermediaries, thus reducing transaction costs. Nevertheless, despite their functionality, they have a speculative tendency, generating a kind of alternative (and less sophisticated) financial market that is still beginning to attract other, more traditional institutional players such as banks and corporations. Another critical factor of the crypto market (and another ironic twist on the decentralization promise of DLT) is that several entities of the token economy are centrally managed (such as example, wallet providers, exchanges, and custodians).

---

[24] European Commission, Proposal for a Regulation of The European Parliament and of the Council on Markets in Crypto-assets, and amending Directive (EU) 2019/1937, COM/2020/593 final.

[25] See Chap. 6 by Vieira dos Santos and Maia in this book.

It is unclear to what extent the attention afforded to the economic value of tokens does not soak all the interest in protocol development and adoption.

1.4 The Law in Blockchain

Blockchain is a technology and an economic reality. As a communication technology, it is used to progress human interaction; as an economic reality, it is a production of value. In both ways, blockchain is a social phenomenon and, therefore, a legal one too.

Where is the law in blockchain? Put: whenever there is an agreement between two or more people. People download the protocol and run it, accepting that action to participate in a joint enterprise. By participating in the enterprise, they subject themselves to the protocol rules of governance and functioning. If a malicious intent wrongs their legitimate expectations, they have the right to claim damages from those who wronged them. Suppose they were not wronged, and there were legitimate expectations that they were, by downloading the protocol, participating in a risky activity. In that case, they are at their peril without solid legal protection.

Users of the network transact with one another. Transactions are subject to legal rules and principles. They use a smart contract, but a smart contract is a way of executing the transaction and not a legal agreement subject to specific rules that are "alegal". After all, programmable code is a language written by people and can represent declarations and agreements.[26] Concerning interactions with people, download a wallet, open an account in an exchange, download a phone app to the custody of their tokens: these are all legal agreements, subject to contract rules. As for tokens, they are digital representations of value that, if similar to other legal categories of value (e.g., securities, property titles, identity marks), must abide by the applicable regulatory standards. Apart from El Salvador and the Central African Republic, crypto-assets do not enjoy legal tender status in any other jurisdiction. However, that does not make them, per se, forbidden payment methods (legally, they are just equivalent to banter; if people agree to pay and be paid in crypto, there is, in general, no consequence).[27]

The picture above is simplistic, I admit, and does not consider some complex situations. For example, regarding the rights and duties of participants in the network: if a miner fails to mine a block and insert a transaction from a user, can he/she claim damages against the miner for loss of profit? If the protocol's code has a bug, and that bug enables the running of illegal transactions, which is to be responsible? The protocol's creator, the foundation/company that created and takes care of the protocol's maintenance? The freelance programmers and developers that dedicate their time (often subsidized by grants from the founding companies, foundations, or protocol investors)? What about voting and participation? Do I have a right to

---

[26] See Chap. 8 by Ferreira in this book.
[27] Unless, of course, crypto payments are forbidden in that specific jurisdiction.

transparency by staking and voting in protocol governance? In the case of a hard fork that I did not agree with, can I claim damages against people that supported it? Moreover, what is the rules (contract rules, corporate rules) applicable to the situation?

The question of applicable law and jurisdiction is a matter of international private law that holds an essential set of rules for deciding the law applicable to the relation and the forum for exercising its rights. An article by Andrew Dickinson is elucidative in showing that, despite the difficulties, it is possible to discern what are the rules applicable to a situation connected with a blockchain.[28] The legal questions surrounding the rights and duties of all actors in the protocol—miners, holders, and members of decentralized autonomous organizations (DAOs)—must be seen in the light of the specific jurisdictional rules applicable to this scenario, regarding contractual private and legitimate expectations, as rights and duties applicable to joint ventures and other forms of informal or irregular association.[29]

It is also essential to consider the differences between protocols when assessing the legal implications of the rights and duties of users. For example, in terms of expectations and agreements: the majority of the most-valuable protocols by market cap are enterprise or corporate-backed protocols with marketing structures in place. Individual users rarely deal directly with the protocol; instead, they deal with third-party companies (or, if you do not want to avoid the irony of it all: intermediaries) that are directly connected with the protocol. Most users will be legally protected (at least on paper) by the agreements signed with service providers. As for users (e.g., companies, institutions, and States) that wish to participate in the protocol, they are also protected given that they relate directly with the protocol's backers: the foundations and companies that created the protocol and have an interest in it to succeed. Expectations are agreed upon and legally established ex-ante, with a clear identification of the parties and their corresponding contractual duties. Therefore, the risks of governance decentralization are somewhat mitigated.

The bottom line is that when duly considered in their meaning and upon understanding the true extent of practical implications, the challenges of "decentralization" and automated processing (or smart contracts) are not as complex as initially believed. Decentralization is not an antonym to law: it has existed for a long time, particularly in political systems such as federalism and feudalism. It means that power (of deciding, commanding, or ruling) is not (or not entirely) in the hands of one central entity or group but instead distributed by many points.[30] It is a complicated system, for it involves two or more power centers relating and executing powers in a separate or coordinated matter. Nevertheless, decentralization is never absolute; some degree of centralization exists for efficacy purposes. Federalism is, once again, a good way of explaining this when we consider the power balance established between the two centers of political authority (the center and the periphery—the federation and the federated States). What matters is the covenant, the pact, the agreement—the

---

[28] Dickinson 2019, 94–136.
[29] Rolo 2019, 33–87, and Wright 2021, 152–176.
[30] See the question in Bodó and Giannopoulou 2019.

1 Blockchain and the Law: Setting the Floor   11

protocol—that establishes the basis and structure of power. As soon as this is identified and settled, legal connections for establishing rights, duties, and liability will flow.

Blockchain is a private endeavor, a market created by people and managed by people, even if based on digital platforms with automated mechanisms of interaction in place. People establish the rules through agreement and within the space of autonomy the law provides for them. If there is a problem with the code, in terms of how something is impossible to achieve (for example, data privacy), it is necessary for the code to be amended, which can happen.

The situation with Bitcoin is the most difficult one from this perspective. The anonymity of Nakamoto, together with the lack of a centralized structure behind the protocol and the multiple actors that are, in one way or another, invested in it, makes it very challenging to find adequate legal protection for users and to understand the rights and duties of each actor within the network. There is some form of legal practice, call it a digital-community custom at play, in the way the members of the Bitcoin community established procedures for updating the network.[31] Still, the process is somewhat erratic and informal. Nevertheless, given the fact that in terms of market adoption, most people deal (and shall deal) with intermediaries and, therefore, not connect to the network directly, and given the liability between the user and the intermediary, I do not consider the situation very problematic, from a realistic standpoint.

1.5 The Law for Blockchain

Blockchain is thus a legal phenomenon because the law covers and regulates blockchain and its developments and interactions. Another question is if this "coverage" (or regulation) is adequate. I will try to provide some ideas in this section.

The law receives social and technical phenomena under its terms. The operation of law, however, is complex, with a significant role afforded to hermeneutics and interpretation. The more open the meaning of rules is, the bigger the capacity of the law to adapt to new phenomena. One excellent example is the definition of what constitutes security under EU law, which is open enough to include any manifestation of value that works as an investment instrument.[32]

Is law "adaptive" enough to deal with blockchain and its developments? We have seen that private law rules can deal with several situations of blockchain technology. However, the space of autonomy of private law ends, and some things that tamper with the greater adoption of blockchain are not permitted. For example, it is generally not legal to establish a negotiation system of financial instruments in a decentralized blockchain. It is also illegal (or at least devoid of legal effects) to represent a property title from a house, car, or corporate store as a cryptographic token; it is necessary to

---

[31] The book by Jonathan Bier shows this perfectly, see Bier 2021.
[32] See Ferrari 2020, 325–342.

execute the act through a prescribed form of legal act (like a public deed, regarding real estate acquisitions). Although some regulation establishes incorporation criteria for people exercising an economic activity with "virtual assets" to prevent money laundering, it is unclear whether credit institutions can add tokens to their portfolio or not.[33] The tripartite and artificial division of token classification also makes it difficult to understand how to treat gains from the sale of crypto assets from a tax law perspective.[34] Furthermore, it is unclear whether blockchain can serve as a system for the valid verification of identities.[35]

The most significant legal challenge of blockchain is not dogmatic but practical: DLTs are global protocols operated by several people, spread around the world, with transactions being done in a peer-to-peer way among different jurisdictions. Identities are cryptographically drawn, and pseudonyms are used, which makes it difficult sometimes to locate and enforce. Global cooperation and technical expertise of law enforcement and courts are necessary. Global recognition of crypto assets in a form of legal category could also help with financial and tax regulation.

In sum, legal rules must enable digitalization for the blockchain revolution to succeed (and the digitalization of value to occur). There are already rules over automated decision-making and processing, but not much concerning digitalizing identities, certificates, public deeds, or other forms of authenticated data. Moreover, most of these property certificates are issued by public institutions or accredited actors, such as notaries. To think of a "decentralized" State apparatus is fascinating from a technical side. Public administration is *the* centralized social institution per excellence, even if it can be divided into departments or different administrations (e.g., regional, municipal, federal.). In the end, it is the public sector (the State, in its multiple formations) that controls the protocol, even if having a blockchain can bring gains in efficiency and transparency.

1.6 Conclusions

Four years after the hype, blockchain continues to deliver at its slow pace. The modern world—particularly the modern communication and media—expects immediate and impactful results to be pleased. Once these are over, attention moves to the next big thing. That is the cycle of modern life if we want to be realistic.

Although there is already an extensive and profound body of research, the fact is that we are still at the beginning. One thing is to cover a project, a proof-of-concept, a promise, or something potential, and another is to cover an existing,

---

[33] See Directive (EU) 2018/843 of the European Parliament and of the Council of 30 May 2018 amending Directive (EU) 2015/849 on the prevention of the use of the financial system for the purposes of money laundering or terrorist financing and amending Directives 2009/138/EC and 2013/36/EU, PE/72/2017/REV/1.

[34] See Chap. 4 by Carmo in this book.

[35] EU Blockchain Observatory and Forum 2020a.

active, and established reality. Blockchain adoption is growing, as is blockchain development, which brings new faces and features that, on the one hand, help simplify comprehension and impact of the initial challenges (governance decentralization, for a start). However, on the other hand, it also brings new forms of complexity (e.g., yearn finance, DeFi, and staking operations). There is also more regulation and awareness, as well as some mainstream adoption of ideas, such as token issuing, NFTs, and central bank digital currencies.

There is still much to research, much to consider, and much to legislate, many litigations to come. It is not clear how blockchain will impact the real world of human interaction, if we will be all participating in the decentralized economy or if we will allow others to do it for us. I confess I suspect it will be the latter, given that general computational literacy is very low for the population to participate in these systems actively. Moreover, the power of States and the capacity of multinational companies and legacy actors should not be underestimated. Digital revolutions have appeared and gone, and people's appetite for security may trump the benefits of decentralized digital governance.

Lex cryptographica may, one day, become a field of law, but as a sub-form of private law or regulatory law, as the set of general and specific rules applicable to commercial and social interactions made through the use of DLT protocols and crypto assets. What is essential to take note of is that blockchain and DLT developments are constantly evolving and will play a role in the communication and value-transfer infrastructure of the future; traditional contractual and commercial rules may apply to blockchain interactions, be them among users, network founders, miners, and the like; enforcement of local rules can sometimes prove challenging and can be inadequate for it may risk tampering with the creative side of the phenomenon; new rules should be enacted in some cases, but the most significant change that has to come is cultural, concerning computational literacy and understanding of how digital technologies develop. If there are notes I would like to suggest regarding a blockchain framework and the law, these are the ones.

References

Alexander L (2019) Law and Politics: What Is Their Relation? University of San Diego Legal Studies Research Paper Series no. 19–412
Allot P (1999) The Concept of International Law. European Journal of International Law Volume 10, 31–50
Aristotle (2002) Edition of Nicomachean Ethics. Oxford University Press, Oxford
Bier J (2021) The Blocksize War. Self-Published
Bodó B, Giannopoulou A (2019) The Logics of Technology Decentralization—The Case of Distributed Ledger Technologies. Institute for Information Law Research Paper No. 2019-02. https://papers.ssrn.com/sol3/papers.cfm?abstract_id=3330590, accessed 31 October 2022
Cerar M (2009) The Relationship Between Law and Politics. Annual Survey of International & Comparative Law Volume 15
Ors C (2021) What's Next for Blockchain? 3rd Generation Platforms. https://medium.com/web3labs/whats-next-for-blockchain-3rd-generation-platforms-a26f34da4d59, accessed

31 October 2022
Copeland T (2020) Steem Vs Tron: The Rebellion Against a Cryptocurrency Empire. Decrypt. https://decrypt.co/38050/steem-steemit-tron-justin-sun-cryptocurrency-war, accessed 31 October 2022
De Fillipi P, Wright A (2018) Blockchain and the Law: The Rule of Code. Harvard University Press. Harvard
Dickinson A (2019) Cryptocurrencies and the Conflict of Laws. In: Fox D, Green S (eds) Cryptocurrencies in Public and Private Law. Oxford University Press, Oxford
EU Blockchain Observatory and Forum (2020a) Legal and Regulatory Framework of Blockchains and Smart Contracts. https://www.eublockchainforum.eu/sites/default/files/reports/report_legal_v1.0.pdf., accessed October 31, 2022
EU Blockchain Observatory and Forum (2020b) Governance of and With Blockchains. https://www.blockchain4europe.eu/wp-content/uploads/2021/05/report_governance_v1.0_0.pdf, accessed October 31, 2022
Ferrari V (2020) The regulation of crypto-assets in the EU – investment and payment tokens under the radar. Maastricht Journal of European and Comparative Law Volume 27: 325–342
Fink M (2018) Blockchain Regulation and Governance in Europe. Cambridge University Press, Cambridge
Fox D, Green S (eds) (2019) Cryptocurrencies in Public and Private Law. Oxford University Press, Oxford
Hacker P et al (eds) (2019) Regulating Blockchain: Techno-Social and Legal Challenges. Oxford University Press, Oxford
Hughes G (1968) Rules, Policy, and Decision Making. The Yale Law Journal Vol. 77, No. 3, 411–439
Kapasi H (2021) Blockchain technology can change the world, and not just via crypto. Cointelegraph, https://cointelegraph.com/news/blockchain-technology-can-change-the-world-and-not-just-via-crypto, accessed 31 October 2022
Kelsen H (1967) Pure Theory of Law. University of California Press, Oakland
Luhmann N (2004) Law as a Social System. Oxford University Press, Oxford
Magnusson W (2020) Blockchain Democracy and the Rule of the Crowd. Cambridge University Press, Cambridge
Mangabeira Unger R (1976) Law in Modern Society. The Free Press
Meier J, Schluppi B (2019) The DAO Hack and the Living Law of Blockchains. In: Molin-Kränzlin A et al. (eds) Digitalisierung—Gesellschaft—Recht. Dike
Narayanan A et al (2016) Bitcoin and Cryptocurrency Technologies: A Comprehensive Introduction. Princeton University Press, Princeton
Rauchs M et al (2018) Distributed Ledger Technology Systems: A Conceptual Framework. Cambridge Centre for Alternative Finance. https://papers.ssrn.com/sol3/papers.cfm?abstract_id=3230013, accessed 31 October 2022
Raz J (1979) The Authority of Law. Oxford University Press, Oxford
Rolo A (2019) Challenges In the Legal Qualification of Decentralized Autonomous Organizations (Daos): The Rise of The Crypto-Partnership? Revista de Direito e Tecnologia, Volume 1. 33–87
Spilka D (2021) How Blockchain Will Change the Way We Work, Play and Stay Healthy in the Future https://www.nasdaq.com/articles/how-blockchain-will-change-the-way-we-work-play-and-stay-healthy-in-the-future-2021-08-26, accessed 31 October 2022
Wright A (2021) The Rise of Decentralized Autonomous Organizations: Opportunities and Challenges. Stanford Journal of Blockchain and the Law, Volume 4.2: 152–176

Other Documents

Directive (EU) 2018/843 of the European Parliament and of the Council of 30 May 2018 amending Directive (EU) 2015/849 on the prevention of the use of the financial system for the purposes of money laundering or terrorist financing and amending Directives 2009/138/EC and 2013/36/EU, PE/72/2017/REV/1

European Commission, proposal for a Regulation of The European Parliament and of the Council on Markets in Crypto-assets, and amending Directive (EU) 2019/1937, COM/2020/593 final

Chapter 2
Blockchain and Elections: Opportunity and Peril

Nelson Rosario

Contents

2.1	Introduction	17
2.2	Elections	19
2.3	Blockchains	22
2.4	Combining Blockchains and Elections	25
2.5	Conclusion	27
References		28

Abstract This chapter discusses the opportunities and perils associated with adopting blockchain-based systems to solve longstanding problems in elections. The chapter covers some of the historical problems in elections and how they were solved, as well as some of the benefits of blockchain technologies. Several attempts to apply blockchain solutions to the election domain are analyzed and potential paths forward are suggested.

Keywords Blockchain · Elections · Crypto · Distributed ledger technology · Election administration · Election systems · Political science

2.1 Introduction

As the old saying goes: it is the counting of votes that matters, not the votes themselves. This cynical saying is generally attributed to the former Soviet dictator Joseph Stalin, though it is likely Stalin never said this as the provenance of the saying is contested, but even so the sentiment crosses generations.[1,2]

N. Rosario (✉)
Rosario Tech Law, LLC; Illinois Tech, Chicago-Kent College of Law, 435 W. Diversey Pkwy, Suite 1A, Chicago, IL 60614, USA
e-mail: nelson@rosariotechlaw.com

[1] https://www.snopes.com/fact-check/stalin-vote-count-quote/.

[2] https://www.politifact.com/factchecks/2019/mar/27/viral-image/no-joseph-stalin-didnt-say-statement-about-electio/.

The provenance of our votes, and the assurance that comes with knowing that your vote *counted* the way that you intended is what builds trust in elections and their results. In this environment, many people see blockchain technology as a solution to build out a digital system of voting. As one example, former US Presidential candidate Andrew Yang famously suggested moving voting in the USA to a mobile-based blockchain system.[3] His bold assertion stated *"It is 100% technically possible to have fraud-proof voting on our mobile phones today using the blockchain. This would revolutionize true democracy and increase participation to include all Americans—those without smartphones could use the legacy system and lines would be very short."*[4] Technological solutions do not exist in a vacuum and technology that is deployed to solve problems in the election world are no different.

Blockchains are an effective solution to the "double-spend" problem[5] for cryptocurrencies and other digital assets. Blockchains are not, however, a good solution to electronic voting let alone voting in general. Technological solutions to problems in the election space must take into consideration the political environment, the cultural environment, the legal environment, and the technological environment.[6] Without additional changes to an electoral system blockchain technology as it is commonly deployed adds little value to an electoral system.

According to IBM, a blockchain is *"a shared, immutable ledger that facilitates the process of recording transactions and tracking assets in a business network."*[7] The IBM definition is useful in that its focus is on transactions and assets that exist in a *business* network. Indeed, transactions of an economic nature is where blockchains as we know them today began. The very first blockchain of economic significance was the Bitcoin blockchain. As IBM claims, the benefits of blockchain are greater trust, greater security, and more efficiencies.[8]

Although blockchains may be an effective solution for the "double-spend" problem, as applied to cryptocurrencies and other digital assets, they are not, on their own, a good solution for electronic voting. The reality is that blockchain technology requires a lot of additional changes to an electoral system to add value.

This chapter begins by looking at the intersection of blockchain technology and elections. The relevant background on both blockchain and elections is explored covering their history, characteristics, as well as some benefits and drawbacks. Then the chapter turns to the reasons for the growing interest in using blockchain for elections. In particular, the current state of the art regarding blockchain and elections

---

[3] https://www.forbes.com/sites/billybambrough/2019/08/27/andrew-yang-2020-us-presidential-election-issue/.

[4] *Id*.

[5] The "double-spend problem" is a flaw in digital cash protocols that allows the same digital token to be spent more than once.

[6] "Can Blockchains Safeguard Elections?"—Nelson M. Rosario, https://inthemesh.com/archive/can-blockchains-safeguard-elections/.

[7] "Blockchain defined" https://www.ibm.com/topics/what-is-blockchain.

[8] "Benefits of Blockchain" https://www.ibm.com/topics/what-is-blockchain.

is examined, and the potential benefits and drawbacks of using this technology for voting systems is explored.

The key requirements for elections and the technological deficiencies of blockchains to solve elections are explained to demonstrate to the reader that blockchains are not a good solution for electronic voting. By the end of this chapter, readers will have a better understanding of how technology, in particular blockchain technology, may or may not be used in elections and the potential implications of using this technology moving forward.

2.2 Elections

An election is one of the key elements of a political system, and instituting any changes to an election process requires a nuanced understanding beyond what many people may realize. While individuals have little difficulty envisioning the act of voting on the day of the election, the massive administrative structure necessary to facilitate such an event is often overlooked. Election systems have developed over time, resulting in the complex systems we see today, with historical antecedents in civilizations like Greece and Rome. For example, Athens around the 5th century BCE used a direct democracy-style system, while the Roman Republic employed elections to select a variety of officials across its government.

The evolution of election systems over time has led to a range of different types of elections, including general elections, local elections, and referendums, each with their own unique requirements, opportunities, and challenges. The early democratic systems were primarily based on direct democracy, in which citizens voted directly on political matters. However, modern elections have become more complex and incorporate various forms of representative democracy and electoral systems such as the first-past-the-post system,[9] proportional representation,[10] and ranked-choice voting.[11] Additionally, historical and cultural precedents may dictate election norms around the world that impact actual election administration.

Elections are typically characterized by a set of common characteristics related to suffrage, secret ballots, competition, timing, and administration. For example, there must be some measure of suffrage for citizens to participate in the election process. The ballots used to cast votes should be kept secret to protect individuals' privacy

---

[9] "First-past-the-post voting" is a system known as winner-takes-all, where voters indicate their preference for a single candidate, and the candidate with the most votes wins. See, generally, https://en.wikipedia.org/wiki/First-past-the-post_voting.

[10] "Proportional representation" is an electoral system in which voters choose a political party rather than a candidate, and seats in parliament are allocated to each party in proportion to the number of votes they receive. See, generally, https://en.wikipedia.org/wiki/Proportional_representation.

[11] "Ranked-choice voting" is a voting system that allows voters to rank candidates in order of preference, and then uses those rankings to determine a winner. Advocates believe that the system allows for greater voter choice and ensures that the winning candidate has the support of a majority of voters. See, *generally*, https://hls.harvard.edu/today/ranked-choice-voting-explained/.

as well as mitigate political reprisals. There should be free and fair competition amongst candidates and political parties, and elections should occur at regular intervals. Finally, it is crucial that the election process be administered in a transparent and accountable manner to ensure the public's confidence in the electoral system.

It is essential to understand the complexity of election systems, including their historical context and development, to appreciate the challenges involved in ensuring that they are conducted fairly and transparently. While many individuals may take for granted the fundamental tenets of democracy such as voting, it is critical that lawyers, government regulators and academics remain attentive to the intricate details of the electoral process to ensure that it operates efficiently and effectively, ideally promoting democratic values and ideals.

As already discussed, elections have long been considered a cornerstone of democratic societies, providing citizens with a mechanism to elect their leaders and actively participate in the political process. Elections provide a variety of very important benefits to society. One of the key benefits of elections in modern democracies is the representation they afford to citizens. Through the democratic process of voting, citizens are empowered to select officials whose policies and positions align with their values, beliefs, and preferences. As such, elections serve as a critical means of ensuring that the government is truly representative of the people it serves.

In addition to representation, elections also play a critical role in promoting accountability among elected officials. By providing constituents with the power to select their leaders, elections create a system of checks and balances that ensures elected officials are accountable to the people they serve. This accountability ensures that elected officials remain responsive to the needs of their constituents, promoting good governance and reducing the likelihood of corruption and abuse of power.

Elections also play a vital role in maintaining political stability by providing a peaceful means of transferring power. Regularly held elections may ensure that political power is not concentrated in the hands of a single group or individual, and may reduce the likelihood of political instability, unrest, or violence. The peaceful transition of power through elections also helps promote confidence in the political process and create a sense of continuity and stability within the government.

Finally, elections may confer legitimacy upon elected officials, as they are chosen by the people to represent their interests. The ideal way to achieve this is by providing a transparent and fair process for selecting leaders, and in that way elections help ensure that elected officials are perceived as legitimate representatives of the people they serve. This legitimacy is critical for promoting stability, building public trust in the government, and promoting effective governance. Ultimately, elections are a critical component of democratic societies, playing a vital role in promoting representation, accountability, stability, and legitimacy.

An election is an essential cornerstone of democratic societies, but it is not without its shortcomings. For one thing, elections are costly to administer, with extensive expenses related to voter registration, ballot production on the front-end, and vote counting on the back-end, making it difficult for some countries to hold elections in a regular and fair manner. This can lead to low voter turnout, which can dilute the legitimacy of election results and weaken the democratic process, making it challenging to ensure that elected officials represent the people they serve.

Additionally, the elections are susceptible to manipulation, fraud, voter suppression, and external interference that can jeopardize the integrity of the democratic process, leading to a decline in public trust and even sparking political instability, unrest, and violence.[12]

Overall, while an election is an indispensable aspect of democratic societies, its disadvantages, including cost, low voter turnout, and susceptibility to manipulation, must be carefully considered when designing and implementing electoral systems. It is vital to ensure that the benefits of elections are not outweighed by their drawbacks, and that the electoral process remains equitable, transparent, and accountable to the people it serves. Technological solutions are no panacea to these practical considerations.

Indeed, all elections are subject to a common set of concerns. Technology, such as blockchain technology, has the potential to revolutionize the electoral process, but it also raises concerns about the security and integrity of elections. One of the most significant concerns is voter fraud, which occurs when individuals or organizations manipulate the electoral process through activities such as ballot stuffing, vote buying, or impersonating voters. The use of technology in the electoral process has the potential to mitigate these concerns, with solutions such as biometric identification, blockchain technology, and other security measures that can help ensure the accuracy and integrity of the vote. These are legitimate concerns, but the prevalence of ballot stuffing,[13] vote buying,[14] and impersonating voters[15] must also be considered.

Another major concern associated with the use of technology in elections is voter suppression. Efforts to systematically prevent or discourage certain groups of people from voting, often targeting marginalized populations, can undermine the legitimacy of election results and weaken the democratic process. While technology can be used to increase access to voting, such as mobile voting or online voting, it also raises concerns about security and potential manipulation, making it essential to strike a balance between accessibility and security.

Finally, the use of technology in elections has also raised concerns about external interference. Attempts by foreign actors to influence election outcomes, through methods like hacking, disinformation campaigns, or financial support for favored candidates, can undermine the integrity of the democratic process and lead to political instability.[16] It is essential to ensure that appropriate safeguards are in place to protect

---

[12] See, for example, alleged Russian interference in the USA Presidential Election of 2016: https://en.wikipedia.org/wiki/Mueller_report and https://archive.org/details/MuellerReportVolume1Searchable/Mueller%20Report%20Volume%201%20Searchable/.

[13] https://www.scientificamerican.com/article/what-does-A-crooked-election-look-like/.

[14] https://theconversation.com/a-third-of-indonesian-voters-bribed-during-election-how-and-why-100166.

[15] https://www.businessinsider.com/voter-election-fraud-statistics-rare-president-biden-trump-2020-2020-11.

[16] See footnote 12.

against external interference and that the electoral process remains transparent and accountable to the people it serves.

In order to address the concerns detailed above, policy makers have adopted a variety of approaches. One such area of focus for policymakers and election officials is the implementation of voter identification laws. Requiring voters to present identification to verify their eligibility may help prevent voter fraud and ensure the integrity of the vote. However, it is essential to ensure that these laws do not disproportionately impact marginalized communities or violate individuals' privacy rights.

Another key area of focus is election monitoring. Independent observers or organizations can monitor elections to ensure transparency and identify potential irregularities. The use of technology, such as blockchain, can help increase transparency and improve the accuracy of election monitoring.[17] However, it is also essential to ensure that these monitoring efforts do not violate individuals' privacy rights or create unnecessary barriers to the electoral process.

Campaign finance regulations are another critical component of the electoral process. These laws govern the funding of political campaigns to prevent corruption and undue influence. The use of technology, such as online fundraising platforms, can help increase transparency and accountability in campaign finance. However, it is essential to ensure that these platforms do not facilitate illegal or unethical behavior.

Investing in more secure voting systems, improved voter registration processes, and better training for election officials is essential to strengthen the electoral process. The use of technology, such as, potentially, biometric identification and blockchain, can help increase the security and accuracy of the vote. However, it is essential to ensure that these systems are accessible to all citizens and do not create unnecessary barriers to voting.

Increasing transparency is critical to maintaining the integrity of the electoral process. This can be achieved through the use of technology, such as open engagement with the technology community and real-time election monitoring. However, it is essential to ensure that these transparency efforts do not violate individuals' privacy rights or create unnecessary barriers to the electoral process. By balancing the benefits of technology with the need for security, privacy, and transparency, we may ensure that the electoral process remains fair, accessible, and accountable to all citizens.

2.3 Blockchains

Blockchain technology has its roots in the development of the Bitcoin cryptocurrency in 2009.[18] While initially viewed as a fringe technology, blockchain has since gained widespread attention for its potential to revolutionize various industries, including finance, healthcare, and elections. The technology has evolved over time, with various

---

[17] Election monitoring is one way in which that election-focused blockchain companies have attempted to improve electoral systems around the world.

[18] See, generally, https://bitcoin.org/en/bitcoin-paper.

implementations of supposed blockchain technology, such as private and public blockchains, being developed.

The characteristics of blockchain technology include a varying degree of decentralization, immutability, and security. Decentralization refers to the fact that blockchain transactions are stored across a distributed network of computers rather than in a central location. Immutability refers to the fact that once a transaction is recorded on the blockchain, it is difficult to alter or delete it. Security is achieved through the use of cryptographic algorithms that ensure that only authorized parties can effectuate and modify blockchain transactions.

While blockchain technology has many potential benefits, including increased trust, security, and efficiency, it also has several drawbacks. These drawbacks include the potential for high energy consumption, scalability issues, and the complexity of implementing blockchain systems. Additionally, the use of blockchain technology in elections has raised concerns about privacy, security, and accessibility, making it essential to carefully consider the potential benefits and drawbacks of blockchain technology in the electoral process.

What exactly is a blockchain though? A blockchain is a new way of organizing information that relies upon existing technologies that are well understood. One such definition is that a blockchain is a *"data structure [that] is an ordered, back-linked list of blocks of transactions"*.[19] Note the focus on transactions again. Unpacking this definition a bit more, here our data structure[20] is a means to organize our data in an ordered fashion where each block of data is linked together and contains a set of transactions per block. To further explain this idea, *"[b]locks are linked 'back,' each referring to the previous block in the chain. The blockchain is often visualized as a vertical stack, with blocks layered on top of each other and the first block serving as the foundation of the stack."*[21] Through the use of a cryptographic hash function[22] each block in the chain has a unique fingerprint used to identify it. So, each block in the chain has a fingerprint that identifies the previous block in the chain *"[s]o each block not only tells us the where the value of the previous block was, but it also contains a digest of that value, which allows us to verify that the value hasn't been changed."*[23]

There is a distinction to be made between public and private blockchains. Public blockchains are those blockchains that are permissionless and decentralized. The most prominent examples of public blockchains are the Bitcoin blockchain and the Ethereum blockchain. When we say that these blockchains are permissionless it means that anyone anywhere in the world can join these blockchain networks if

---

[19] Antonopoulos 2014, p. 159.

[20] A data structure is simply a way to organize information stored in computer. See: https://www.merriam-webster.com/dictionary/data%20structure.

[21] *Id.*

[22] A cryptographic hash function is a mathematical function that can take any amount of input data and once that data is run through the function an output with expected characteristics is generated. The key feature of a hash function is that any change at all in the input data will generate a different hash. For more explanation see: https://mathworld.wolfram.com/HashFunction.html.

[23] Narayanan et al. 2016, p. 11.

they have the correct software necessary to participate. When we say that these networks are decentralized, we mean that transaction data is distributed across a network of computers, with no central authority or intermediary controlling the flow of information. This ensures that no single entity can manipulate or alter the data, ideally providing increased security and transparency.

Conversely, private blockchains need not guarantee permissionless nor decentralization. The advantages to not guaranteeing those characteristics are that private blockchain deployments allow for complete control over who may participate in the network as well as how information flows through the network. As such, private blockchain deployments are often favored by large corporations who require more certainty in their data management practices as well as by government entities. Additionally, since private blockchains are not permissionless or decentralized there is a question of whether they are in fact blockchains at all. In particular, detractors of private blockchains often believe that a small number of entities may retain control of code to run the blockchain network, as well as reverse transactions or delete transaction entries that would have been considered legitimate in a public blockchain network.

So, what problems do blockchains actually solve? One of the primary issues that blockchains address is the problem of trust. Traditional digital systems, such as online payment platforms, require users to trust a third party to facilitate transactions. This trust can be easily broken if the third party is hacked or decides to act dishonestly.[24] With blockchains, transactions may be verified and recorded across a decentralized network of computers, eliminating the need for a trusted intermediary. This increases the security and transparency of digital transactions, providing users with greater control and confidence in their interactions. Additionally, blockchains solve the problem of data tampering by creating a hard to alter record of transactions that provides evidence when it has been altered or deleted. This ensures a higher degree of integrity in digital data, mitigating unauthorized modifications and providing a potentially more reliable source of truth for all parties involved.

While blockchains are a powerful and innovative technology, they also create several problems that must be addressed. One of the primary issues with blockchains is their complexity. Blockchain systems are often difficult to understand and implement, requiring specialized knowledge and expertise. This complexity can create barriers to entry, limiting the number of organizations and individuals who can effectively utilize the technology. Additionally, blockchains can be resource-intensive, requiring significant computing power and energy consumption to operate. This can create environmental concerns and limit the scalability of blockchain systems. Finally, the anonymity of blockchain transactions can create challenges for law enforcement and regulatory agencies, as it can be difficult to identify and hold accountable those who engage in illegal activities on blockchain networks. Overall,

---

[24] See, for example, the irregularities and problems associated with Wirecard in Germany: https://en.wikipedia.org/wiki/Wirecard_scandal#Causes_of_downfall, and the partially successful hack of the Bangladeshi Central Bank in 2016: https://www.reuters.com/article/us-cyber-heist-philippines-idUSKCN0YA0CH.

while blockchains offer many benefits, it is essential to carefully consider and address the challenges and drawbacks associated with the technology.

2.4 Combining Blockchains and Elections

Governments may be interested in combining blockchain technology and elections for several reasons. One of the primary benefits of using blockchain for elections is the assumed increased security and transparency it provides. By creating a supposed immutable record of transactions, blockchain technology provides a promise to prevent voter fraud and tampering, ensuring that election results are accurate and trustworthy. Additionally, blockchain may help improve the efficiency of the electoral process by streamlining voter registration, ballot distribution, and vote counting. This may reduce the potential for errors and delays, providing faster and more reliable election results. Blockchain technology may be an excellent fit for something like voter registration, because whether someone is registered to vote or not is a different problem than whether someone has voted or not.

Another reason governments may be interested in combining blockchain and elections is to increase voter participation. By using mobile-based blockchain systems, governments can make it easier and more convenient for citizens to vote, particularly those who may face barriers to traditional voting methods, such as physical disabilities, geographic isolation, or mobility issues. This can help increase voter turnout and ensure that all citizens have a voice in the political process. Overall, by leveraging the benefits of blockchain technology, governments may be able to create more secure, efficient, and inclusive electoral systems, enhancing the legitimacy and effectiveness of democratic governance.

While blockchain technology has the potential to improve the security and efficiency of the electoral process, there are several reasons why governments may be hesitant to combine blockchain and elections. One of the primary concerns is the complexity and cost of implementing blockchain systems. Blockchain technology requires specialized knowledge and expertise to implement effectively, which can create barriers to entry for governments and electoral officials. Additionally, the cost of implementing blockchain systems can be prohibitive, particularly for developing countries or those with limited resources. This can make it challenging to ensure that all citizens have access to secure and reliable electoral systems.

Another concern with combining blockchain and elections is the potential for technical issues and vulnerabilities. While blockchain technology is generally secure, it is not immune to hacking or other forms of cyber-attack. This can create the potential for external interference in the electoral process, which can undermine the legitimacy of election results and weaken public trust in the democratic process. Additionally, the anonymity of blockchain transactions can make it difficult to ensure that only eligible voters are participating in the election, which can create challenges for voter identification and fraud prevention. Overall, while blockchain technology has many potential benefits, governments must carefully consider the potential drawbacks and challenges before implementing blockchain in the electoral process.

All around the world, elections have been under the threat of cyber-attack on their election infrastructure for some time now. In 2017, The International Institute for Democracy and Electoral Assistance (International IDEA) hosted a roundtable discussion on cybersecurity and elections. Three points were discussed during the event. Firstly, it is essential to understand the potential risks and vulnerabilities of the electoral process to cyber-attacks, and to develop strategies to mitigate these risks. Secondly, there is a need for greater collaboration and coordination among government agencies, electoral officials, and other stakeholders to ensure effective cybersecurity measures are implemented. Finally, it is crucial to maintain public trust and confidence in the electoral process by ensuring that election results are secure and transparent, and that citizens have access to accurate information about the electoral process and cybersecurity risks. The roundtable discussion highlighted the importance of addressing cybersecurity concerns in the electoral process to ensure the integrity and legitimacy of democratic elections.[25]

It is in this environment that governments everywhere turn to the appeal of blockchain technology to be used in elections. There have been calls in the United States to implement blockchain solutions for elections from a variety of entities across the political spectrum.[26] The concern is real and understandable and has led to a few attempts to utilize blockchain technology for election purposes.

For example, in 2018 in Sierra Leone, a Swiss organization named Agora used their proprietary private blockchain system to help audit the results of the presidential election in Sierra Leone.[27] As stated by the Sierra Leone government, the Agora team was part of a set of international observers to add transparency to the process. By all accounts, Agora's involvement was successful and did not endanger the election, but Agora did not facilitate the election itself. In fact, Agora's involvement caused some confusion among the press as to what exactly their role was.[28]

The most successful attempt to use blockchain in elections has to be what the company Voatz has been able to do thus far. Voatz is a private company based in the United States focused on building a mobile elections platform that utilizes a blockchain.[29] Voatz has successfully conducted a variety of public elections in the United States.[30] Individuals are allowed to request the ability to vote on their mobile

---

[25] "Cybersecurity and Elections: An International IDEA Round-table summary" by Peter Wolf https://www.idea.int/news-media/news/cybersecurity-and-elections-international-idea-round-table-summary.

[26] "How blockchain could improve election transparency" https://www.brookings.edu/blog/techtank/2018/05/30/how-blockchain-could-improve-election-transparency/ and "Blockchain and Election Integrity" https://www.newamerica.org/digital-impact-governance-initiative/blockchain-trust-accelerator/around-the-blockchain-blog/blockchain-and-election-integrity/.

[27] https://www.coindesk.com/markets/2018/03/08/sierra-leone-secretly-holds-first-blockchain-audited-presidential-vote/.

[28] https://futurism.com/sierra-leone-election-blockchain-agora.

[29] https://voatz.com/faq/.

[30] https://voatz.com/.

device, they verify their identity using biometrics, and they cast their vote on the same mobile device.[31] The votes are secured and audited via a private blockchain operated by Voatz.[32]

Not all experts agree that blockchain based voting is a good idea though. A paper published in November of 2020 argued that blockchain based voting could result in changes to elections without even realizing the changes occurred.[33] Firstly, blockchains are not a solution for the complex challenges of the voting process, as they do not solve issues such as voter fraud and vote buying. Secondly, blockchain technology is still in its early stages and requires significant development to be effective in addressing election-related challenges. Finally, there are other more straightforward and effective ways to address the challenges of the electoral process, such as improving voter education and training for poll workers.[34] A key distinction between the error allowance for commerce versus elections is that *"[f]or elections there is no insurance or recourse against a failure of democracy,"* Rivest says. *"There is no means to 'make voters whole again' after a compromised election."*[35]

2.5 Conclusion

Elections are the bedrock of modern democracy, and our modern world is a technological one with problems unimaginable to the Ancient Greeks and Romans. Given the stakes involved and the ever-growing threat of cyber-attacks on election systems around the world it is understandable that governments wish to employ seemingly more secure methods of voting. That said, not every technological tool is necessarily a good fit for every problem.

Blockchain technology allows for the creation of append-only ledgers of transaction data that are tamper-evident, tamper-resistant, public and auditable, and enable new forms of human cooperation. Voting is a kind of double-spend problem where you do not wish to allow people to vote more than once. The siren song of the blockchain fixing elections is forgivable, but governments must take care. Quick fixes rarely are quick or fixes. The blockchain may be a good use of creating digital cash, but that does not mean it can obviate all the problems facing modern elections today. Additional research into how to integrate blockchain systems into existing election systems is needed before those election systems are thrown out and replaced with a ledger system not designed to solve election problems.

---

[31] *Id.*
[32] *Id.*
[33] https://www.csail.mit.edu/news/mit-experts-no-dont-use-blockchain-vote.
[34] *Id.*
[35] *Id.*

References

Antonopoulos A (2014) Mastering Bitcoin. O'Reilly Media, Sebastopol, California
Narayanan A et al. (2016) Bitcoin and Cryptocurrency Technologies. Princeton University Press, Princeton, New Jersey

Nelson Rosario, Founder and Partner at Rosario Tech Law, LLC and Adjunct Law Professor at Illinois Tech, Chicago-Kent College of Law, 435 W. Diversey Pkwy, Suite 1A, 60614 Chicago, IL, USA

Chapter 3
What Are the Limits of Blockchain? Considerations on the Use of Blockchain in Transitional Justice Processes

Renato Gomide M. de Almeida

Contents

3.1	Introduction	30
3.2	What Is Blockchain?	31
3.3	Transitional Justice and the Effectiveness of Using Blockchain	34
	3.3.1 Truth Commissions	36
	3.3.2 Prosecution	39
	3.3.3 Restoring Democratic Institutions	40
	3.3.4 Reparations	42
	3.3.5 Reconciliation	45
3.4	Conclusion	46
References		48

Abstract Blockchain technology has the capacity to change how people live, communicate with each other, acquire products, store information, and exchange data. Such a technology—one that is able to perform in different segments and which is being called a general-purpose technology, just like the internet,—can be useful in transitional justice processes. This chapter intends to analyze the use of blockchain technology as a new component within transitional justice processes (see Sect. 3.3.2), understand how blockchain would behave as part of a transitional justice scenario (see generally Sect. 3.3), by considering a state-of-the-art situation (see Sect. 3.3.5). More specifically, the chapter analyzes how blockchain could be incorporated into important aspects of transitional justice: criminal prosecutions, truth commissions, reparations, the restoration of democratic institutions, and reconciliation.

Keywords Transitional justice · Blockchain · Access to justice · Rule of law

---

R. Gomide M. de Almeida (✉)
Eça de Queiroz Street, 279, Ap 4 Vila Mariana, São Paulo, SP 04011-031, Brazil
e-mail: renato.g.m.almeida@gmail.com

3.1 Introduction

The use of technologies used to be restricted to State actors, which are commonly the principal offenders in transitional justice and human rights situations, which, in turn, leads to the monopolization of these instruments in the hands of offenders. However, in recent years such monopolization has become less intense and new technologies, such as radars, digital media, apps, drones, and open-source mapping, have become much more common among non-state actors. It allows the civic population to better record and transmit violations, helping to prepare more substantial and better procedures against them. These technologies have the power to alter the balance of power between perpetrators and transitional justice defenders since such technologies that once were exclusively in the hands of a few are now part of the civil society reality, enabling information about abuses to be gathered and better analyzed.[1]

Transitional justice continually makes use of new technologies to increase its efficiency in different fields. From identifying and recovering missing people,[2] to gathering and categorizing information for the truth-seeking process.[3] Several are the positive impacts in introducing new procedures and technologies into transitional justice, with the idea to improve the quality and time of its operations.

In that sense, adherence to a general-purpose technology[4] (GPT) such as blockchain may be essential to dealing with transitional justice issues. The lack of support from governments and other actors may jeopardize transitional justice projects that constantly struggle to find proper personnel and funding for their operations. Therefore, the adaptation of existing technologies to the transitional justice reality would be useful, especially when such technologies are not a single purpose one,[5] since it is a more reasonable and cost-effective solution to deal with such issues.

The present chapter intends to create new perspectives regarding how using blockchain technology can help to solve current problems in transitional justice procedures. If understood as a general-purpose technology, with auditable transactions, data storage capabilities and the ability to work in different fields, blockchain technology might be quite suitable to be adapted for transitional justice needs. In blockchain, what is transacted is a digital representation of the information (a digital token), enabling the recording and transaction of any data.

The chapter is divided into two segments. The first (Sect. 3.2) intends to briefly establish the concept of blockchain, while the second (Sect. 3.3) will set out some of the principal elements of transitional justice using Jeremy Sarkin's theory of the five pillars of transitional justice and propose new procedures using blockchain as a catalyzer of these pillars' efficiency.

---

[1] Pham and Aronson 2019.

[2] Sarkin 2017.

[3] Gavshon and Gorur 2019.

[4] General-purpose technologies (GPT) can influence several sectors simultaneously, like electricity and the internet. Werbach 2018, p. 72.

[5] Piracés 2018.

The conclusion demonstrates that blockchain technology has enormous potential in different fields, including transitional justice. Its capability to operate in various areas, such as data sharing, digital identification, fundraising and distribution of money and goods, as well as the time-stamping of documents in a faster, secure, and public way, has significant value for transitional justice scenarios. However, one must bear in mind that blockchain has technological limitations and consider how it might affect transitional justice programs. In the end, the only way to properly analyze how blockchain can assist in a transitional justice project is by understanding each case individually, and by examining the characteristics of blockchain technology with its limitations in a specific scenario.

3.2 What Is Blockchain?

First, it is important to define what constitutes blockchain technology, especially its technological aspects, and then, move forward to deciding under what conditions its application would be possible. There are several potential applications in the context of Transitional Justice (TJ) processes, such as the digital identification of populations and digital registration of documents (i.e., land registries).

The first time that blockchain technology was presented to the world was in Satoshi Nakamoto's paper on Bitcoin.[6] This paper became the core model for the development of blockchain technology since it was the first-ever paper introducing this technology into practical transactional situations; it was completely virtual, and could be trusted in a trustless (i.e., decentralized) environment.[7] Satoshi's article is based on previous studies on the Byzantine Generals Problem[8] and digital time-stamping.[9] The first study intended to understand how computer science can operate in adversarial environments, dealing with conflicting information, making an allusion to the Byzantine Generals Problem of trust situations (how the lack of trust in distributed systems can be overcome, and unanimity reached) while the second study on time-stamping deals primarily with the notion of a blockchain-like state where digital data could be efficiently time-stamped using cryptographic technology.[10]

In a nutshell, blockchain works as a "peer-to-peer" version of electronic transactions using a "block"[11] system, thereby removing all forms of intermediary or central authority from its financial transactions, and is, therefore, described as anarchic governance.[12] The program works on an open-source system, allowing any

---

[6] Nakamoto 2009.
[7] Users can trust the system without having to trust in other users or a third party.
[8] Lamport et al. 1982.
[9] Harber and Stornetta 1991; Bayer et al. 1993.
[10] Rauchs et al. 2018, p. 15.
[11] The definition of Andreas Antonopoulos is: "A block is a container data structure that aggregates transactions for inclusion in the public ledger, the blockchain". Antonopoulos 2017, p. 196.
[12] Rauchs et al. 2018, pp. 72–74.

participant to join and operate in the network. The only limitations are the technical ability, equipment capability and internet efficiency of the user.[13]

The transactions are pieces of cryptographic information recorded in "blocks" which are organized in the form of a chain (hence the name "blockchain"). As transactions develop, a mathematical problem arises leading to a cryptographic number, the "fingerprint" of the block.[14] This number is called a "hash", and each block contains such a digital sign.[15] This process can be done in several ways, but the most common one is called Proof-of-Work[16] (PoW), where participants compete, using their CPU power to solve the mathematical problems first, and then creating the block. The winner is usually financially compensated with cryptocurrencies. The validation of transactions is by a consensus mechanism, where after the block creation, all the others consent about its existence. Blocks can be rejected, and their "creator" is not compensated for their effort. Therefore, most honest users must control the system if they want it to work. Users must blindly trust the network, believing that the other users are doing the same.[17,18]

One other important aspect is that the hash of a block contains the information of the previous blocks on the chain to which it belongs. Therefore, changing a previous block will consequently change (recalculate), the forward blocks on the chain.[19] However, the cost of this kind of change is massive computational and power energy that would not compensate for the effort. As the longer the chain of blocks is, more difficult it is to alter the history of the blockchain,[20] making such technology practically immutable.[21] Therefore, the more transactions occur, the more reliable the system becomes. The network agrees not only about what happened but also about the sequence of events.

Anonymity is another characteristic that ensures security for the users. To access a blockchain system and to transact with others, users must have a pair of private-public keys, where both keys have identification numbers and are protected with encryption. The private-key gives the user private access to the system (and contains personal information such as login and password), and activates the public-key, which functions as the user's identification.

---

[13] Ibid., p. 73.

[14] Ibid., p. 502.

[15] Antonopoulos 2017, p. 197.

[16] According to the Ethereum Whitepaper the Proof-of-Work "was a breakthrough in the space because it simultaneously solved two problems. First, it provided a simple and moderately effective consensus algorithm, allowing nodes in the network to collectively agree on a set of canonical updates to the state of the Bitcoin ledger. Second, it provided a mechanism for allowing free entry into the consensus, while simultaneously preventing sybil attacks". Buterin 2021.

[17] It is the resolution of the Byzantine Generals, problem described above. Blind trust in the system ensures that good decisions are made that do not harm the network.

[18] Werbach 2018, p. 501.

[19] Antonopoulos 2017, p. 195.

[20] Ibid., p. 196.

[21] Werbach 2018, p. 503.

When dealing with specific situations, other features can be added to a blockchain platform. There are other attributes that might be crucial when dealing with TJ, such as scenarios, and these have to be considered when designing the architecture of a blockchain system for this purpose.

Smart contracts is a blockchain technology that allows all transactions to occur automatically, depending exclusively on the initiatives of both parties to the transaction.[22] It is a code that automatically executes what was disposed of, via codification.[23] Smart contracts are used in Ethereum and other blockchain systems to eliminate a counterpart risk. So, when one performs an order to buy or transfer goods from someone else in the blockchain, it is unable to retract the order. It is a warranty for the fulfilment of the deal, ensuring more trust in the system's transactions.[24] In that sense, smart contracts can also be programmed under specific conditions, like ensuring an obligation from time to time, for example, and be triggered after a particular event.[25] To activate the smart contract application oracles are used. Oracles are the ones who feed the smart contracts with external information, outside of the blockchain, starting the actions of such contracts that were pre-shaped. Oracles can be either an automated data source or a human.[26] In the TJ scenario, smart contracts could be used to deliver monthly payments to victims and victims' families, avoiding bureaucratic delay by the State often associated with such cases.

Usually, blockchain systems are publicly available and no permission to participate is needed. However, there are other types of blockchain systems that are a better fit for TJ requirements.

When considering TJ, it is worth noting that blockchain technology systems can be divided into two groups: public and private networks. The public one is how Bitcoin blockchain operates, as a publicly available network where anyone can see what is transacted. A private network, on the other hand, is not open to the public and is constituted by an owner who decides who can have access to the system.

Also, the system can determine how users interact with the platform. The division is between permissionless and permissioned systems. Permissionless systems allow any user to be part of the transaction and consensus network, as in the case of the Bitcoin blockchain, where anyone with enough capabilities can integrate and operate within the system. A permissioned network, by contrast, restricts participation in the system to only those with permission granted by the owner of the network to do so.[27] This means that in a permissioned blockchain, only authorized users can

---

[22] For a deeper understanding of smart contracts in the judicial world: Herian 2018.

[23] Fink 2018, p. 670.

[24] There are several definitions of what a smart contract is as well as different applications and variations of smart contracts. Also, smart contracts represent the term "code is law", which means that one is bound by code to fulfil what is disposed of in code. However, it may generate judicial issues in the case of miscoding or bad faith.

[25] Fleuret and Lyons 2020, p. 15.

[26] Werbach 2018, pp. 213–214.

[27] Ølnes et al. 2017.

add blocks to the chain and validate those nodes.[28] Thus, it creates a system with greater accountability that keeps its operations transparent, but one that is restricted to certified users only.

In brief, a blockchain system can be either public or private and be permissionless or permissioned. The Bitcoin blockchain, for example, is a public, permissionless, blockchain system, while a blockchain system that controls the supply chain of a private company, say, may be a private and permissioned one that restricts the nodes to the company's members. In other words, a blockchain will have different characteristics, depending on its purpose.

In a TJ system, different combinations of networks can exist. A private and permissioned blockchain can be used for digital identification matters, with sensitive information secure in a database controlled by one entity. A public and permissionless blockchain, like Bitcoin's, can also exist for a funding and donations platform to support the TJ pillars. There are several possibilities, which can be tailored to requirements and the local context.

Using all the tools mentioned above, blockchain has evolved and is now at the leading edge of a new technological revolution, creating a trustless but trustworthy environment for transactions without the necessity for a central controller party.

Although initially used for financial transactions, the technology can record any type of information, and can be used for various purposes. It is this feature that makes blockchain a general-purpose technology.

3.3 Transitional Justice and the Effectiveness of Using Blockchain

Transitional justice is broadly discussed in the international community. It has also been applied in several situations around the globe, which are the subject of extensive literature and case studies.

The United Nations Secretary-General first published a report in 2004 addressed to the UN Security Council regarding aspects of the rule of law and TJ in post-conflict situations.[29] The report discusses the perceptions of the United Nations (UN) regarding TJ, as well as its objectives, mechanisms, and procedures to address transitional situations effectively.

The UN has refined its understanding of TJ mechanisms, or pillars, via the Guidance Note of the Secretary-General[30] on its approaches to TJ.

---

[28] Although the owner of the system grants access, the protection measures such as consensus, PoW and encryption remain the same. The only difference is that there is a central authority that chooses who can participate in the network and who cannot.

[29] United Nations Secretary-General 2004.

[30] United Nations Secretary-General 2010.

Although there are several definitions of TJ, most authors follow a similar understanding to the one contained in UN reports. In this regard, transitional justice is

> the full range of processes and mechanisms associated with a society's attempt to come to terms with a legacy of large-scale past abuses, in order to ensure accountability, serve justice and achieve reconciliation.[31]

For the UN, transitional justice is, therefore, applicable to past large-scale abuses by legal and psychological mechanisms to overcome such issues. The arrangements stated by the UN are prosecution initiatives, the right of truth, reparations, institutional reform, and national consultations. Furthermore, such processes must be driven by certain guiding principles, including compliance with international norms and standards, awareness of the local political context, minority- and victims-centered approaches, and the involvement of national law enforcement to ensure the rule of law.[32] All elements of TJ must work holistically to be properly implemented. In that sense, blockchain technology can unite all TJ approaches and, at the same time, operate individually in each one of them by being programmed to act in each situation, and choosing the best network condition for each need.

Even though there are different understandings and concepts regarding TJ, its pillars and approaches, a broad consensus of scholars understand that there are four preliminary pillars of TJ, namely prosecution, truth commissions, institutional reform, and reparations.[33] There are several critics of this format of TJ pillars, with extensive literature on the subject[34] which does not fall within the scope of this chapter. Therefore, in the name of objectivity, the five pillars chosen as the object of study are those stated by the UN, with the addition of reconciliation as the fifth pillar.[35]

One constant critique in TJ approaches, however, is that they regularly ensure restorative and retributive justice, with prosecutions and truth commissions, in a Western approach. Such a method undermines the TJ processes, removing the protagonism of the local society to the international/Western community, resulting in local ownership being lost. TJ approaches also fail to address the root causes of the conflict, such as socioeconomic structures and social inequalities.[36] In that sense, a comprehensive system using blockchain technology that ensures the protagonism of the local population, with local leaders integrated into the system, could help to enhance local participation. Also, the possibility of sending financial aid via blockchain to victims could improve local participation by empowering victims economically.

---

[31] Ibid., p. 2.

[32] Ibid.

[33] Villalba 2011, p. 2.

[34] To understand some of the critics of the U.N. understanding and approach to Transitional Justice, please see: Nagy 2008. And Lambourne 2014a.

[35] Classification used by some scholars such as Jeremy Sarkin.

[36] In that sense, authors such as Wendy Lambourne call for another perspective on Transitional Justice, namely "Transformative Justice". For a deeper understanding, please see: Lambourne 2014b.

The intention now is to elucidate what the pillars of TJ are, and then propose new systems using blockchain technology in each TJ mechanism.

3.3.1 Truth Commissions

A Truth Commissions (TC) or Truth and Reconciliation Commissions (TRC) deals with the pursuit of truth in a non-judicial way, which is one of the principal elements when dealing with past atrocities. According to "Truth Seeking," a publication by the International Center for Transitional Justice (ICTJ),[37] the importance of truth relies on the healing process, restoring the dignity of victims and combating impunity and public denial. From this perspective, TCs play an essential role in psychological healing, investigatory work and addressing issues that emerge from conflicts.

TCs are usually constituted by an executive decree or by law and take into consideration the specific characteristics of each country's situation.[38] However, three objectives of truth commissions are fundamental:[39] to investigate and clarify the violent events that have occurred and those that are denied or disputed, including the socioeconomic context that gives rise to the events; to adopt a victim-centered approach, protecting and empowering victims to overcome the problems suffered, and the preparation of a final report detailing the causes of the conflict and recommendations to avoid repetition. TCs do not have the power to prosecute perpetrators; its investigations can be used in criminal tribunals for that purpose by working alongside judicial organs.[40]

Such an essential and decisive organism of TJ has specific characteristics and considerations that must be observed to be as effective as possible.[41] Truth commissions must be transparent and legitimate in the eyes of civil society as well as ensure the participation of nationals in their procedures. They also must be independent, especially from the State, in all forms (economically, personnel, management) to avoid interference from third parties and maintain their credibility.

Truth Commissions can also be Truth and Reconciliation Commissions (TRC), which also pursue reconciliation. Some specialists understand TRC as an organ that does not merely seek the truth of past abuses but understands the role of reconciliation in such contexts. Reconciliation works adds another layer of challenge, which is why only some truth commissions have added the notion of reconciliation to their name and approach, the most well-known example being the South African TRC.[42]

Notwithstanding the above, it is also important to mention unofficial truth projects (UTPs). Such projects resemble TCs in that they promote the historical reconstruction

---

[37] González and Varney 2013, p. 4.

[38] However, in some cases truth commissions can be established outside the States apparatus.

[39] González and Varney 2013, p. 9.

[40] Avruch 2010, p. 34.

[41] González and Varney 2013, pp. 10–12.

[42] Avruch 2010, pp. 39–40.

of past abuses in a specific place and time. However, the main difference between official and unofficial truth commissions is that TCs have the official sanction of the State, whereas UTPs do not.[43] Usually, UTPs derive from the coordination of civil society organizations that tend to replicate the model and goals of TCs,[44] and are often implemented when there is no State initiative or an unwillingness on the part of the State to implement a TC. However, it does not prevent TCs and UTPs from coexisting, or one transmuting into the other with time.[45]

However, against a background of limited human and financial resources, significant outreach policies and structures for public awareness and participation, information gathering may be a challenge in some cases.[46] Blockchain technology can address some of those issues, due to its ease of accessibility, and its ability to create a financial network, and rapidly gather and transact audited information.

Truth Commissions may suffer from limited resources due primarily to the State's pressure and influence. In several cases, States are unwilling to engage in uncovering the truth, and for that reason, TCs may suffer from fewer contributions from States.[47] Although the State fully funds many TCs, as in the case of the Chilean one, others receive funds from abroad, such as the Salvadoran TC, which received funding from both local government and international actors, as did the South African TRC.[48]

Due to their unofficial status UTPs depend on private investment from institutions or civil society to survive and, therefore, have an even greater lack of resources than their officially backed counterparts, the TCs.[49]

One possible solution to this problem is to gather public financial aid by crowdfunding projects based on blockchain. Donors would receive tokens from their operations, as compensation (i.e., discount on products) or, in other cases, securities issued by the company.[50]

Crowdfunding is a fundraising method where many people financially contribute small amounts to a project via the internet and receive rewards from the project.[51]

Nonetheless, in a TJ situation where a TC needs funds to carry out its work properly, initiatives such as crowdfunding are a viable option. Due to the elimination of the mediator party in blockchain systems, the funds invested would go directly to the truth commission's funds. A crowdfunding initiative would also allow broader participation by the global community since anyone with an internet connection and sufficient resources to make a minimal contribution could donate. In a blockchain system, cryptocurrencies could also be incorporated to provide another donation payment system. In return for a donation, the TC could issue a utility token which would, say, give

---

[43] Bickford 2007, p. 1026.
[44] Ibid., p. 1002.
[45] Ibid., pp. 1004–1005.
[46] Villalba 2011, p. 8.
[47] Hayner 2011, p. 217.
[48] Ibid.
[49] Bickford 2007, pp 1028–1029.
[50] Cai 2018, pp. 974 and 985.
[51] To more fully understand crowdfunding issues, please see Belleflamme et al. 2013.

the donor access to documents from the commission that are not yet available to the public, or some other incentive.[52] This proposal is already being studied by organizations such as the OCHA (United Nations Office for the Coordination of Humanitarian Affairs) as a crowdfunding method for humanitarian emergencies.[53]

Another problem caused by a lack of resources is the restricted outreach capacity of TCs that would otherwise enable public participation in truth processes. TCs have an important social role, and social involvement is essential to their work and legitimacy.[54] Public engagement is crucial for truth-seeking, gathering information from the population, ensuring that victims' voices are heard by the community, promoting transparency and public participation and creating the notion of ownership of the TC.[55]

One other challenge faced by TCs is managing data issues, especially with regard to the gathering, analyzing and cataloging of each piece of information. It can present enormous problems, especially nowadays, when new technologies, such as smartphones with video or audio recording capabilities, and social media, can generate and disseminate a fantastic amount of data very rapidly.[56]

For both issues, it is important to state that, in a blockchain system, the transfer of information is extremely fast and, with the use of a smartphone in places with an internet connection, lots of different data points can be collected and audited and be immediately accessed by users.

In a public blockchain system, all the information exchanged can be accessed by anyone with an internet connection and a smartphone, which improves the system's transparency and outreach. Also, the network can be permissioned, meaning that users must ask for permission to trade and insert information into the platform. So, the system controller (that could be the TC itself or another controller) can allow only specific users to interact with the network. It can provide another layer of legitimacy, as well as better control of the information gathered.

In that sense, as the information in the network is confirmed, the nodes (users) could stamp it as valid information. The logic here is similar to a supply chain process, but with pieces of data instead of products. So, as the data flows through the different validation processes, it becomes more legitimate, and since the entire process is publicly available, it is also transparent for civil society. Therefore, cases of misinformation and excessive useless data are controlled and minimized.

---

[52] A collateral issue that may arise is the case of tokens issued by TCs being traded in secondary markets. On the one hand, this could promote the entrance of the TCs into a quasi-financial system (not unlike a stock market). On the other hand, the tokens could be programmed with an expiration date, and be automatically destroyed after a specific date or event.

[53] Salazar et al. 2015.

[54] Bickford 2007, p. 1030.

[55] González and Varney 2013, p. 49.

[56] Gavshon and Gorur 2019, pp. 72–73.

3.3.2 Prosecution

Prosecution of perpetrators, also known as retributive justice, is the most traditional way to bring those responsible for crimes to account.[57] According to the UN Guidance Note,[58] prosecutions aim to ensure accountability for those responsible for committing gross human rights and humanitarian violations, and their being tried according to international standards of human rights and humanitarian law. There is an extensive tradition of international tribunals that prosecute perpetrators of international criminal law. From the Nuremberg trial to the International Criminal Tribunal for the Former Yugoslavia (ICTY) and the International Criminal Tribunal for Rwanda (ICTR).[59] These previous *ad hoc* tribunals were created for persecutory reasons and serve as a foundation for the current international criminal law theory and jurisprudence, shaping current procedure.

Nowadays, the State has the primary jurisdiction to conduct the prosecutions, with the assistance of TJ programs, enhancing and reinforcing judicial capabilities.[60] In the case of States unwilling or unable to prosecute perpetrators, defendants are charged by other means, such as hybrid tribunals[61] or by the International Criminal Court (ICC),[62] which has subsidiary jurisdiction. In these mechanisms, there is an international component of accountability of perpetrators by international standards. However, the intention is the same; to fight impunity.[63]

The ICC has suffered criticisms, though, due to its undermining role as a human security agent for the local populations and Its own *in locus* personnel, and as a complementary organism in retributive and distributive justice, intended to assist the other TJ mechanism and national courts with expertise and infrastructure.

Still, there are also procedural challenges that may arise in prosecuting processes, such as securing pieces of evidence, identification of perpetrators, the protection of victims and witnesses, international cooperation between States and the ICC, preservation of evidence and sharing of information.[64]

Blockchain technology can provide a data share network for specific markets, where all information can be accessed and transacted by all nodes. In a public permissioned system, transactions have an additional degree of trust.

In that sense, issues of securing and sharing information are solved, and, with only specific nodes being able to insert data, information exchanged is controlled by only those entities involved in the prosecution process. Notwithstanding, all data

---

[57] Lambourne 2014b, p. 20.
[58] United Nations Secretary-General 2010, p. 7.
[59] Meron 2006, p. 91.
[60] Investigatory, prosecutorial, judicial, defensive, witness and victim protection and support.
[61] Hybrid courts have a mixed characteristic of being composed of national and international elements, usually operating in the jurisdiction where the crimes occurred: Office Of The United Nations High Commissioner For Human Rights 2008a.
[62] According to Article 17 of the Rome Statute.
[63] Villalba 2011, p. 3.
[64] Ibid., p. 4.

is stored in the network, meaning that no central authority has absolute custody of what is transacted, thereby adding another layer of protection to the system.

The use of blockchain to gather information on victims and witnesses could be of crucial help since organizations with authorization to add blocks in the system could go directly to meet these people to take their statements, with anonymity granted by the network. In other words, such actions could introduce another layer of protection to victims, witnesses, and the local team responsible for collecting such statements.

Blockchain technology could also help the ICC to monitor and assist local personnel and national courts with the exchange of information. Having the ICC as one node in a blockchain system would facilitate the international transfer of data between different national and international groups and promote faster responses to local situations. For example, the collection of a statement of a victim *in locus* would allow immediate access to this data by the ICC, permitting accuracy and agility in the decision-making processes of the ICC as well as those of the authorities and civil organizations to which the Court has access. A pre-established international blockchain network, with international organizations, such as the UN and the ICC, as nodes and other international actors, such as NGOs, could be a solution for faster response times. The network owner could allow the participation of other members for a specific situation, and then remove them later on.

In addition to the above, and as a solution proposed by TCs, the ICC could initiate a crowdfunding project based in blockchain, allowing financial support from any sponsor, and establish the possibility for funding via cryptocurrencies as well. The use of cryptocurrencies could have a positive impact since their use could simplify the funding processes and make it cheaper. The ICC could use such funding support to promote its actions as a complementary organization, as discussed, and also financially support local organizations, such as hybrid courts, TC projects and reparations programs via the Trust Fund for Victims (TFV). Therefore, the ICC could enhance its complementary function as a body pursuing all types of justice and as a human security agent.

As demonstrated, the use of blockchain technology for prosecution purposes in a TJ situation is similar to its use in TC situations, since it intends to improve the efficiency of operational aspects, such as the exchange of information between several actors.

3.3.3 Restoring Democratic Institutions

After periods of violence, usually caused by the State, there is distrust regarding the State and its institutions, especially as regards the preservation of rights. Inn such contexts, certain State institutions could symbolize the State regime and, therefore, should be dismantled or changed to regain public trust.[65]

---

[65] One example is the old DOI/CODI in Brazil, symbol of the Brazilian military dictatorship and torture practices during the regime, scrapped as part of the Brazilian democratization process.

According to the UN Guidance Note to TJ,[66] public institutions that helped to perpetuate violations should be transformed into institutions that sustain peace, human rights, and the rule of law. This pillar is seen as a guarantee of non-repetition, a way to prevent future human rights violations.[67] Mechanisms for non-repetition usually focus on human rights violations (in the justice and security sectors), by using vetting tools or by means of lustration,[68] to lead to more independent institutions,[69] as well as to create a culture of human rights protection, via educational training of security agents and complete educational reform directed to future generations. However, this pillar depends on political will, which can be more difficult when there are personnel at high levels that are accused of committing human rights abuses. Nevertheless, such information on State personnel perpetrators should be gathered for future actions, especially as a naming and shaming device.[70] Notwithstanding the above, blockchain technology can improve government efficacy, making autonomous processes in a decentralized and open way, reducing bureaucracy and combating corruption schemes.

The use of blockchain technology in Governments is a subject of broad debate nowadays. Uses of the technology in several fields has the potential to improve public bureaucracy, creating accountability in administrative decisions taken.

Blockchain could be used not only to promote transparency in decisions made by States, but it could also help in other TJ processes associated with the restoration of democratic institutions, in vetting processes, and in education.

All vetting processes and other structural reforms of public organs could be made accountable to an oversight body that would have its procedures publicly available in an open-source blockchain system. Also, all the decisions made would be time-stamped and validated, in digital format, and stored on the network. In that way, all decisions of public organs would be publicly audited via blockchain by an oversight institution.

Regarding the creation of educational programs focused on international human rights and humanitarian law standards for public employees as well as for future generations, blockchain technology can be helpful by monitoring an educational chronogram. A human rights educational system, based on international standards, and powered by blockchain and smart contracts, could be developed in these situations. Employees would log into the system with their identification (or digital ID), and lessons would be based on modules that are available via smart contracts, according to the user's frequency or examination for which they are sitting, thereby giving the authorities better control over the educational programs of government employees and students within the education system.

---

[66] United Nations Secretary-General 2010, p. 9.

[67] Villalba 2011, p. 8.

[68] Vetting and lustration are processes that are designed to remove from office public servants who have been implicated in human rights abuses, normally as part of a previous regime. See De Greiff 2007.

[69] Sarkin 2016, p. 327.

[70] Ibid.

The measures outlined above are deeply politicized. In other words, there must be a political will to put these measures into practice, regain public trust, and promote accountability. Blockchain has the advantage of functioning on the procedural spectrum, providing enhanced transparency, and greater economy.

Blockchain can also be helpful in other areas of government, such as in the fight against corruption,[71] elections (via tamper-proof voting records and smart contracts)[72] or land titles.[73] In dealing with these issues, the immutability and cryptography of blockchain, combined with an open-access platform, can allow a transparent record-keeping process for governments. The implication is that governments would also have better control over digital ID records, storing documents and data transactions between public organs, financial resources and their traceability (especially in the development of stablecoins based on the national currency), criminal records, and passport data.[74] Such areas could be managed through eGovernment applications, which would necessitate the use of more technological tools in governments.

3.3.4 Reparations

Another essential pillar in TJ is reparations. It is not only presented in several international norms but also is considered an international principle by the UN General Assembly via Resolution 60/147 of 2006.[75] Awarding reparations to victims of gross violations is one of the core principles of TJ and international law and should therefore be considered. The State's responsibility to individuals also upholds such a pillar, as well as the perpetrator's responsibility to repair its victims.[76] Therefore, States and individuals should be responsible for their actions or omissions and the harm that they have caused.

Some types of reparations can be linked to the other pillars. In any case, reparations should always be proportional to the harm suffered. The different types of reparations are: restitution (of the victim's previous conditions), compensation (of any economically assessable damage suffered by the victim), rehabilitation (relating to any medical and psychological problems that the victim may have as a result of the harm caused), satisfaction (in the form of truth-seeking, memorialization, an official apology, or judicial or administrative sanctions) and guarantees of non-repetition

---

[71] To understand more about blockchain in the fight against corruption, please see: Aggarwal and Floridi 2019.

[72] To understand more about the e-Voting system, please see: Lyons et al. 2018.

[73] To understand more about land titles via blockchain, please see Eder 2019, and Kshetri 2017.

[74] Ølnes et al. 2017, p. 357.

[75] United Nations General Assembly 2006.

[76] Villalba 2011, p. 5.

(by means of institutional reform of the State apparatus, primarily in the justice and security sectors).[77]

Despite the State's obligation to promote their adequacy and effectiveness, however, reparations programs still face some challenges. They include the outreach of the reparations program ("completeness"),[78] compliance of the reparations program to court judgments to avoid undue enrichment of victims, proper enforce of the reparations program, public participation (especially participation by the victims), and issues regarding collective reparations vis-à-vis individual reparations.[79]

Reparations tend to be broader in scope, since they deal not only with individualized compensation measures but encompass several rights violated that materialize in the personal, collective, civic, as well as political, economic, social, and cultural spheres in psychological and concrete ways.

Issues regarding which violations will be contemplated and the types of reparations that will be granted need to be taken into consideration. However, such problems are generally discussed with political bias and also represent a challenge to foreign actors that monitor TJ processes.

In this area, too, blockchain technology applications can help to create a comprehensive reparations system, one that can assist with delivering aid to victims and track their needs.

Due to the transparency and traceability characteristic of blockchain technology, reparations—especially financial reparations—can be tracked to ensure that the correct amount is paid to the correct victim. Notwithstanding, such capabilities can also be used to monitor health products and other goods destined for specific communities and victims, via a production chain concept.[80] The use of blockchain enables an almost instantaneous monitorization of products via PoW and tamper-proofing of the data transacted, thus ensuring the proper delivery of aid products to victims and avoiding losses. The increased transparency in production chains set up using blockchain technology has the dual benefit of ensuring the proper delivery of aid products to victims, and enhanced accountability for companies that purchase merchandise from human rights violators (e.g., blood diamonds from Sierra Leone).

Regarding land restitution, blockchain technology can provide a new way to register land. Even when land restitution is not possible, States could provide government securities or vouchers as a form of compensation for the loss of the land, which would create a secondary market of such securities.[81] If these securities were tokenized, this secondary securities market could be regulated, but at the same time would introduce compensated victims into the financial system and make the trade of tokens an attractive asset for them.

---

[77] United Nations General Assembly 2006, pp. 7–9.
[78] Office Of The United Nations High Commissioner For Human Rights 2008b, p. 15.
[79] Villalba 2011, p. 6.
[80] Kshetri 2017.
[81] Roht-Arriaza 2013, p. 21.

One of the principal characteristics of blockchain technology is to ensure economic inclusivity for unbanked people[82] and victims of the financial economy, making it easy to effect payments for victims, refugees and displaced people.[83] Due to blockchain specificities, transactions have a smaller cost, which will enable the increase of national and international remittances, enabling victims or victims' families to send and receive money from the reparations programs, without extra cost hurdles. Including victims in such a financial system would increase their sense of empowerment, especially within their own societies, since financial independence from other members of that society can be achieved, especially among women, who are often financially dependent.

Regarding identification and access of victims to reparations programs, blockchain technology can store victims' identifications via digital IDs, and use biometric identity systems (i.e., digital and iris scans)[84] in places where smartphones or internet access is scarce. With digital identification, victims can be easily identified, which would give them access to essential health and education services that usually require some form of identification. Also, it would help to improve government data in the transparent delivery of effective public services and would measure the development progress of a specific community accurately. One such project is Building Blocks by the World Food Programme (WFP)[85] in association with the UNHCR Syrian refugee camps in Jordan and Pakistan.[86]

A system like this could be integrated into a worldwide donation procedure based on cryptocurrency, which would make it possible for anyone to donate to reparations programs. Management of such an infrastructure can be done by the ICC, via the TFV (Trust Fund for Victims) and States. An integrated international approach to deliver reparations could be a positive solution to help States to afford reparations measures.

Reparations management is expensive for governments,[87] due to the operating costs involved in reparations delivery. Blockchain technology could help reduce costs by configuring a smart contract linked to each victim's public key and triggered by oracles. In compensation reparations, for example, the amount can be satisfactorily transferred to each victim. Smart contracts are able to automate all types of transfers and can be programmed to transfer a specific amount at specified intervals for a specific duration to each victim. Smart contracts can be used as a tool of control and ensure that no corruption occurs, since the program works automatically without external intervention.

---

[82] Maupin 2017.

[83] Rella 2019.

[84] With this method, each identity is provided with unicity (no two people will have the same identifier characteristics) and singularity (each person will have one identifier) by the technology.

[85] The WFP also promotes an acceleration program for several start-ups that work in disrupting hunger. Please see: World Food Programme 2020.

[86] Wang and De Filippi 2020.

[87] Hayner 2011, p. 180.

3.3.5 Reconciliation

The UN Guidance Note has National Consultations as its fifth pillar,[88] which states the importance of elaborating an action model tailored to the particularities of each country. This approach has come under academic criticism, since it might lead to the narrowing of the ambit of transitional justice.[89] Reconciliation is regarded as a desired achievement or goal in TJ. With that in mind, reconciliation as a fifth pillar means that the TJ mechanisms should not only aim for reconciliation as a consequence of its approaches but should actively use reconciliatory tools for better results.

Reconciliation is understood as a long-term social process through which psychosocial healing takes place, but which needs the actors to play an active role in forming new relations of peaceful coexistence,[90] with which every other TJ pillar must engage.[91] A transitional environment can effectively promote the context needed to pursue reconciliation.

Issues of truth-seeking, justice and accountability, socioeconomic equality, forgiveness, and dialogue between antagonists are essential to foster reconciliation, as is the outreach of actions to encompass the entire population of the region concerned.[92]

Despite all these issues, reconciliation is deeply dependent on contextual structure, which means that each case needs specific measures that will efficiently apply to the situation at hand.

Understanding reconciliation as TJ's fifth pillar means that it should be at the center of every TJ process. The movement of reconciliation is the movement of TJ, where the use of the other mechanisms must be connected to a common purpose. Reconciliation is not seen as a final objective but rather part of the TJ process. However, including reconciliation as the central pillar of TJ means that reconciliation is not only a process but also an end in itself.

The other pillars of TJ can, however, work separately while not aiming specifically at a reconciliatory process, which is likely to lead to fragile and brief periods of peace.[93] Reconciliation is of a complexity that needs different approaches to succeed.

Blockchain technology can develop a reconciliatory umbrella network, where all TJ mechanisms can exchange data between themselves, in a faster and secure way, to understand how to achieve the common ground required for reconciliation properly.

Blockchain technology can also assist in reconciliation indirectly by acting on the other pillars. All the potential uses of blockchain listed above will, indirectly, help in the reconciliatory process, if their approaches aim at reconciliation.

---

[88] United Nations Secretary-General 2010, p. 9.
[89] Nagy 2008 and Lambourne 2014a.
[90] Avruch 2010 and Aiken 2014.
[91] In that sense Nevin Aiken proposes contact, dialogue, promotion of truth, justice, and the amelioration of socioeconomic inequalities. Aiken 2014, p. 57.
[92] Ibid.
[93] Skaar 2012.

Therefore, the objective of the blockchain network to assist in reconciliation relies on two aspects. The first is indirect, where blockchain acts in other TJ pillars, improving their effectiveness, but with the condition that these pillars have a built-in reconciliatory approach.

The second aspect is the creation of a comprehensive system, based on an umbrella perspective, driven by a blockchain network, where the information of different sectors of TJ can be exchanged between the nodes. Building a completely interchangeable TJ data share network for specific markets will foster reconciliation because it will reunite and distribute information from various TJ fronts, allowing the fast, secure and decentralized exchange of data between the mechanisms.

So, in a state-of-the-art situation, a broad network powered by a public permissioned blockchain system would allow every node (represented by the TJ pillars) to communicate and transact information between themselves. Only nodes with permission would be able to input transactions into the network, but all information would be publicly available.

3.4 Conclusion

This chapter has analyzed how blockchain technology could be related to transitional justice processes, verifying each of the five TJ pillars in a state-of-the-art blockchain situation.

Blockchain technology creates new paradigms for the transaction of information in a decentralized, instantaneous, and secure way. It has created new types of money and assets in borderless environments. It has also allowed a more thorough audit of goods in the production chain, ensuring better standards within the entire ecosystem. At the same time, several governments have been focusing on its potential to modernize the public sector and make it more efficient.

The characteristics of blockchain technology: decentralized control, a shared view of the truth, cross-border collaboration, auditing of data, and the exchange of information/value through digital tokens, combined with its possible public or private, and permissionless or permissioned, network designs could be of great assistance in more sensitive fields, such as transitional justice, especially when the government is viewed as not trustworthy.[94] Using blockchain technology would mean that several processes could be implemented without the need of government machinery, thereby enhancing the transparency of its procedures.

Yet, transitional justice is a field of international human rights and humanitarian law that deals with the past. It works in very complex conditions which, in many cases, lack proper funding for the proper project implementation. Also, there are issues of trust and belonging among nationals and victims. The path towards achieving a reconciliation is a winding one and demands a lot of energy, planning, political

---

[94] Werbach 2018, p. 77.

will, and resources. As a volatile environment, TJ cannot afford mistakes in its undertakings, since it could lead to the return, or worsening, of hostilities.

However, as mentioned before, there is not one single approach to all TJ cases. Blockchain devices, which can adequately adapt to several situations and serve multiple purposes, are suitable for dealing with different demands in a faster, secure, and transparent way.

Although new, blockchain technology has a lot to prove. The potential of blockchain technology to increase effectiveness indifferent areas, and also underpin other prominent technologies, such as the Internet of Things (IoT), or Artificial Intelligence (AI), makes it a very appealing option for several situations. Having a system that does not solely rely on the government itself, but also integrates the international community and the civic population, in a shared view of the truth, is an excellent idea.

In a state-of-the-art blockchain concept, as analyzed in this chapter, the technology demonstrates that it would be extremely beneficial to enhancing efficiency in a comprehensive and holistic TJ approach, making reconciliation a more manageable process. Inclusion of different society sectors and the international community, with the use of an integrated umbrella blockchain system that encompasses all TJ pillars and other essential features can be a reality with blockchain technology. Faster and secure data transactions between permissioned nodes, in a publicly available system, can enhance transparency of processes and improve their outreach. Each TJ pillar can also have its own blockchain system incorporating its specific needs and characteristics. Since each piece of information can be tokenized, the tokens can be freely transacted between the nodes and networks.

Although blockchain is a technology that enhances the transaction and storage of auditable data, its use to empower local populations and victims—providing access to and participation in TJ processes and according people their economic, social, and cultural rights—can change the socio-political and economic relationships present in post-conflict societies.

It is important, however, to consider the challenges that implementing blockchain systems face. There are technological hurdles to overcome in regions where there is a lack of proper internet connectivity or equipment, as well as the existing legal limitations in some countries. It is also worth mentioning that due to is novelty and complexity, blockchain technology is still extremely expensive.

In TJ situations, awareness of these issues is amplified due to their complex environments, and it is well known that errors could cost the lives of innocent people. Issues of mistrust between international actors and States, lack of public participation, lack of political will, underfunding, and other problems add more complexity to processes. Therefore, one must take into thorough consideration how positive can blockchain be and whether other technologies would be more suitable in any given situation. Questions of viability and value are crucial in determining whether blockchain technology will be the best choice in concrete situations.[95]

---

[95] Lindman et al. 2020.

This chapter does not intend to impose blockchain as a full-package TJ solution, but rather proposes this ground-breaking technology as one of several tools that TJ could use to implement, to best effect, mechanisms for reconciliatory processes that result in lasting peace. As such, despite the challenges blockchain faces, it is a technology worthy of serious consideration.

References

Aggarwal N, Floridi L (2019) The Opportunities and Challenges of Blockchain in the Fight Against Government Corruption. 19th General Activity Report of the Council of Europe Group of States against Corruption (GRECO). GRECO 82. March 2019

Aiken NT (2014) Rethinking Reconciliation in Divided Societies. A social learning theory of transitional justice. In: Buckley-Zistel S et al. (eds) Transitional Justice Theories. Routledge

Antonopoulos AM (2017) Mastering Bitcoin. Programming the Open Blockchain, 2nd edn. O'Reilly

Avruch K (2010) Truth and Reconciliation Commissions: Problems in Transitional Justice and the Reconstruction of Identity. Transcultural Psychiatry

Ayed AB (2017) A Conceptual Secure Blockchain-Based Electronic Voting System. International Journal of Network Security & Its Applications (IJNSA). Vol. 9. No 3

Balasco LM (2013) The International Criminal Court as a Human Security Agent. Praxis: The Fletcher Journal of Human Security. XXVIII. 46. January 2013

Bayer D et al. (1993) Improving the Efficiency and Reliability of Digital Time-Stamping. In: Capocelli R, De Santis A, Vaccaro U (eds) Sequences II: Methods in Communication Security, and Computer Science. Springer, New York

Belleflamme P et al. (2013) Crowdfunding: Tapping the right crowd. Journal of Business Venturing

Bickford L (2007) Unofficial Truth Projects. Humans Right Quarterly. Vol. 29. No. 4. https://doi.org/10.1353/hrq.2007.0036

Brounéus K (2003) Reconciliation. Theory and Practice for Development Cooperation

Buterin V (2021) Ethereum Whitepaper https://ethereum.org/en/whitepaper/#a-next-generation-smart-contract-and-decentralized-application-platform. Accessed 30 May 2021

Cai CW (2018) Disruption of financial intermediation by FinTech: a review on crowdfunding and blockchain. Accounting and Finance, Vol 58, Issue 4

Chinkin C (2007) The Protection of Economic, Social and Cultural Rights Post-Conflict. Report Commissioned by the Office of the High Commissioner for Human Rights (OHCHR)

De Greiff P (2007) Vetting and Transitional Justice. In: De Greiff P, Mayer-Rieckh A (eds) Justice as Prevention: Vetting Public Employees in Transitional Societies. Social Science Research Council

Eder G (2019) Digital Transformation: Blockchain and Land Titles. Vienna International Development network. 2019 OECD Global Anti-Corruption & Integrity Forum

Fink M (2018) Blockchains: Regulating the Unknown. German Law Journal, 19(4)

Fischer M (2011) Transitional Justice and Reconciliation: Theory and Practice. In: Austin B, Fischer M, Giessmann HJ (eds) Advancing Conflict Transformation. The Berghof Handbook II. Barbara Budrich Publishers, Opladen/Framington Hills

Fleuret F, Lyons T (2020) Blockchain and the Future of Digital Assets. The European Union Blockchain Observatory & Forum, February 2020

Gavshon D, Gorur E (2019) Information Overload: How Technology Can Help Convert Raw Data into Rich Information for Transitional Justice Processes. International Journal of Transitional Justice, Vol 13

González E, Varney H (2013) Truth Seeking. Elements of Creating an Effective Truth Commission. International Center for Transitional Justice (New York). Amnesty Commission of the Ministry of Justice of Brazil (Brasília)

Harber S, Stornetta WS (1991) How to Time-Stamp a Digital Document. Journal of Cryptology, Vol. 3, N° 2
Hayner PB (2011) Unspeakable Truths: Transitional Justice and the Challenge of Truth Commissions, 2nd edn. Routledge, New York/London
Herian R (2018) Legal Recognition of Blockchain Registries and Smart Contracts
Jacobovitz O (2016) Blockchain for Identity Management. Technical Report #16-02. The Lynne and William Frankel Center for Computer Science Department of Computer Science, Ben-Gurion University
Ko V, Verity A (2018) Blockchain for the Humanitarian Sector: Future Opportunities. Digital Humanitarian Network (DH)
Kshetri N (2017) Will blockchain emerge as a tool to break the poverty chain in the Global South? Third World Quarterly, 2017. Vol. 38, N° 8, 1710-1732. Doi: https://doi.org/10.1080/01436597.2017.1298438
Lambourne W (2014a) What are the Pillars of Transitional Justice? The United Nations, Civil Society and the Justice Cascade in Burundi. Macquarie Law Journal. Vol 13
Lambourne W (2014b) Transformative justice, reconciliation and peacebuilding. In: Buckley-Zistel S et al. (eds) Transitional Justice Theories. Routledge
Lamport L et al. (1982) ACM Transactions on Programming Languages and Systems. Vol 4. No. 3
Lindman J et al. (2020) The uncertain promise of blockchain for government. OECD Working Papers on Public Governance. No 43
Lyons T et al. (2018) Blockchain for Government and Public Services. The European Union Blockchain Observatory & Forum
Maupin J (2017) Blockchain and the G20: Building an Inclusive, Transparent and Accountable Digital Economy. Center for International Governance Innovation, Policy Brief No 101
Meron T (2006) The Humanization of International Law. The Hague Academy of International Law Monographs, Volume 3. Martinus Nijhoff Publishers
Nagy R (2008) Transitional Justice as a Global Project: critical reflection. Third World Quarterly. Vol. 29, No 2. https://doi.org/10.1080/01436590701806848
Nakamoto S (2009) Bitcoin: A Peer-to-Peer Electronic Cash System. Available at https://bitcoin.org/en/bitcoin-paper. Accessed 29 May 2021
Nouwen SMH (2006) "Hybrid Courts": The hybrid category of a new type of internationals crimes courts. Utrecht Law Review, Vol. 2, No 2
Office Of The United Nations High Commissioner For Human Rights (2008a) Rule-Of-Law Tools For Post-Conflict States. Maximizing the Legacy of Hybrid Courts. United Nations. HR/PUB/08/2
Office Of The United Nations High Commissioner For Human Rights (2008b) Rule-Of-Law Tools For Post-Conflict States. Reparations Programmes. United Nations. 2008b. HR/PUB/08/1
Ølnes S et al. (2017) Blockchain in government: Benefits and implications of distributed ledger technology for information sharing. Government Information Quarterly, Vol. 34, Issue 3
Pham PN, Aronson JD (2019) Technology and Transitional Justice. International Journal of Transitional Justice, Vol 13
Piracés E (2018) The Future of Human Rights Technology. A Practitioner's View. In: Land M, Aronson J (eds) New Technologies for Human Rights Law and Practice. Cambridge University Press. https://doi.org/10.1017/9781316838952.013
Rauchs M et al. (2018) Distributed Ledger Technology Systems: A Conceptual Framework. Cambridge Center For Alternative Finance
Rella L (2019) Blockchain Technologies and Remittances: From Financial Inclusion to Correspondent Banking. Frontiers in Blockchain
Roht-Arriaza N (2013) Reparations and Economic, Social, and Cultural Rights. UC Hasting Research Paper no 53. SSRN Electronic Journal. https://doi.org/10.2139/ssrn.2177024
Salazar J et al. (2015) Crowdfunding for Emergencies, United Nations Office for the Coordination of Humanitarian Affairs. Think Brief. OCHA POLICY AND STUDIES SERIES

Sarkin J (2016) Refocusing Transitional Justice to Focus Not Only on the Past, But Also to Concentrate on Ongoing Conflicts and Enduring Human Rights Crises. Journal of International Humanitarian Legal Studies. Vol. 7. No 2. https://doi.org/10.1163/18781527-00702002

Sarkin J (2017) How developments in the science and technology of searching, recovering and identifying the missing/disappeared are positively affecting the rights of victims around the world. Human Remains and Violence: An Interdisciplinary Journal. Vol 3. No 1

Skaar E (2012) Reconciliation in a Transitional Justice Perspective. Transitional Justice Review. Vol. 1. Issue 1. https://doi.org/10.5206/tjr.2012.1.1.4

United Nations General Assembly (2006) Basic Principles and Guidelines on the Right to a Remedy for Victims of Gross Violations of International Human Rights Law and Serious Violations of International Humanitarian Law. Resolution 60/147

United Nations Secretary-General (2004) The Rule of Law and Transitional Justice in Conflict and Post-Conflict Societies. Report of the Secretary-General. S/2004/616

United Nations Secretary-General (2010) United Nations Approach to Transitional Justice. Guidance Note of the Secretary-General

Villalba CS (2011) Transitional Justice: Key Concepts, Processes and Challenges. Briefing Paper. Institute for Democracy and Conflict Resolution (IDCR)

Wang F, De Filippi P (2020) Self-Sovereign Identity in a Globalized World: Credentials-Based Identity Systems as a Driver for Economic Inclusion. Frontiers in Blockchain. DOI: https://doi.org/10.3389/fbloc.2019.00028

Werbach K (2018) The Blockchain and the New Architecture of Trust, Kindle edn. MIT Press

World Food Programme (2020) Building Blocks: Blockchain for Zero Hunger. World Food Programme. Available at https://innovation.wfp.org/project/building-blocks. Accessed 20 May 2021

Zwitter A, Boisse-Despiaux M (2018) Blockchain for humanitarian action and development aid. Journal of International Humanitarian Action, Vol. 3, No 16. https://doi.org/10.1186/s41018-018-0044-5

Chapter 4
Taxing Crypto-Assets—The Portuguese Perspective

Marta Carmo

Contents

4.1	Introduction	52
4.2	Overview of Some Relevant Features of the Portuguese Tax System	53
	4.2.1 The Principle of Tax Legality and Its Importance in the Context of Crypto-assets	53
	4.2.2 The Principle of Legal Certainty and the Binding Rulings	54
4.3	Taxing Crypto-assets under the Corporate Income Tax Code	55
	4.3.1 The Corporate Income Tax Code: A Brief Introduction	55
	4.3.2 Profits with Crypto-assets and the Importance of the Accounting Perspective	56
4.4	Taxing Crypto-assets under the Personal Income Tax Code	58
	4.4.1 The Personal Income Tax Code: A Brief Introduction	58
	4.4.2 The Sale of Cryptocurrencies	59
	4.4.3 Capital Gains?	59
	4.4.4 Capital Income?	60
	4.4.5 Business or Professional Income?	61
	4.4.6 Other Types of Income and Some Practical Considerations	62
4.5	Taxing Crypto-assets under the Value Added Tax Code	63
	4.5.1 The Value Added Tax Code: A Brief Introduction	63
	4.5.2 The European Court of Justice Hedqvist Case	63
	4.5.3 The Sale of Cryptocurrencies	64
	4.5.4 Mining	65
4.6	Conclusions	67
4.7	Afterword	68
References		71

Abstract This chapter analyzes the taxation of crypto-assets in Portugal and the approach taken by the local tax authorities on crypto-assets and selected related topics. After summarizing some relevant features of the Portuguese tax system, the chapter discusses the taxation of profits arising from crypto-assets under the Corporate Income Tax, as well as the importance of the accounting perspective for such taxation. Subsequently, the chapter delves into the topics of Personal Income Tax,

---

M. Carmo (✉)
Banco BPI, Lisbon, Portugal
e-mail: marta.carmo@novalaw.unl.pt

namely vis-à-vis the sale of cryptocurrencies, other types of income and some practical considerations. Finally, taking the Hedqvist case (see Sect. 4.5.2) as its departure point, the tax authority's position on Value Added Tax on the sale of cryptocurrencies and mining is scrutinized.

Keywords Taxation · Principle of tax legality · Crypto-assets qualification

4.1 Introduction

Crypto-assets are complex phenomena from several perspectives, including from the tax one. This chapter explores the Portuguese tax point of view in a non-exhaustive but critical manner, assuming that the reader is not familiar with the Portuguese tax system. However, it takes from granted that the reader is familiar with the main contextual, conceptual and regulatory issues around crypto-assets. Therefore, the chapter tries to provide some insights on the typical tax concerns, namely on the Portuguese tax authority ("TA") position.

Section 4.2 provides a brief overview of some relevant features of the Portuguese tax system, namely the principle of tax legality, the principle of legal certainty and the possibility of any taxpayer to request binding rulings from the TA. This section is relevant in portraying the main uncertainties, struggles and the relative importance of such binding rulings.

Subsequently, the analysis moves into Corporate Income Tax ("CIT"). After a short introduction to this tax, the chapter provides a general framework that we consider to be applicable to companies and other entities subject to CIT that obtain profits derived from crypto-assets, as well as the importance of the accounting standpoint. Curiously, there is no known TA position on this regard.

Afterwards, the chapter covers the Personal Income Tax ("PIT"), providing a quick outline of this tax as well. However, in this case, there's a known TA position (on the sale of cryptocurrencies) which is scrutinized with a critical lens. Then, the chapter discusses other types of income and lays down some thoughts on practical difficulties that may arise in this regard.

Finally, the last section focuses on Value Added Tax ("VAT"), encompassing as well brief introductions to the Portuguese VAT Code and to the well-known European Court of Justice ("ECJ") Hedqvist decision and a review on the two TA binding rules on VAT (one on the sale of cryptocurrencies and a second on mining).

4.2 Overview of Some Relevant Features of the Portuguese Tax System

4.2.1 The Principle of Tax Legality and Its Importance in the Context of Crypto-assets

Crypto-assets are a recent reality. Legal systems should be flexible enough to keep up with the current landscape but in the case of tax law such systems and flexibility should be circumscribed by the principle of legality. In fact, "[t]*he principles of people's sovereignty and separation of powers themselves require that the legal type of tax is discussed and approved by parliaments*".[1] For that reason, the Portuguese Constitution establishes such principle in its article 103.

The principle of legality not only means that tax rules should be approved by the Parliament[2] but also that the essential elements of a tax mentioned in article 103(2) of the Portuguese Constitution ("*Taxes shall be **created by law**, which shall lay down their **applicability, rate, tax benefits and the taxpayers guarantees**"*) should be covered by such approval.[3] The guarantee-based function of this principle "(...) *is associated with the predictability of the amount of tax payable (and thus also with legal certainty)*".[4] Following this rationale, article 11(4) of the Portuguese General Tax Law ("GTL") establishes that "*The gaps resulting from tax rules covered by the Portuguese Parliament's law reserve are not susceptible of analogical integration*".[5] Consequently, when the TA or a Court faces a specific case if the facts fulfil the provision of a statutory tax law, there is a taxable event; if the facts do not correspond to the provision of any statutory tax law (not even by broad interpretation), there is no taxable event (so the taxpayer should not be subject to tax).[6]

Thus, for instance, if one sells a crypto-currency, the resulting income will only be taxed if and only when the tax law covers such type of income. But if the tax law does not cover one crypto-phenomenon, there is no taxable event. To sum up, *nullum tributum sine lege* is very important also in the context of crypto-assets.

---

[1] Dourado 2014, p. 2.

[2] In the Portuguese case, article 165(1)(i) of the Constitution establishes that the Parliament has legislative exclusive competence on the creation of taxes and the fiscal system unless it authorizes the Government to also do so.

[3] Fonseca and Otero 2008, pp. 865–866. Article 8(1) of the Portuguese General Tax Law establishes a provision similar to article 103(2) of the Portuguese Constitution.

[4] Dourado 2007, p. 79—free translation.

[5] Free translation.

[6] Still, one should recall that the Portuguese Statutory Tax Laws use undetermined concepts, grant discretionary powers, assign margins of free appreciation and even allows the conclusion of fiscal contracts (with some boundaries)—e.g., Pires and Calçada Pires 2012, p. 114.

4.2.2 The Principle of Legal Certainty and the Binding Rulings

The principle of legal certainty in tax application (connected with the principle of legality),[7] is a corollary of the rule of law, and it implies the rejection of arbitrariness and the defense of tax transparency, allowing an individual to know in advance to which taxation he or she will be subject and also safeguarding the principle of good faith provided by article 266(2) of the Constitution.[8]

Given the difficulties in qualifying the new realities underlying crypto-assets, especially if we consider that these economic events are not always well understood by decision-makers, this principle gains a greater relevance. Consequently, the importance of binding rulings by the TA stands out since they clarify the interpretation of tax law.

Article 68 of the GTL allows any taxpayer to request a binding ruling on his or her specific tax situation, by stating the relevant facts (as long as they were not subject to a tax audit) and usually by providing a proposal of the respective legal framework.[9] The relevant ruling should be issued within 150 days. The TA cannot proceed subsequently in a direction different from the one in the binding ruling (within the scope of the request), except if in compliance with a judicial decision—see article 68(1), (3), (4) and (14) of the GTL. All binding rulings should be published by electronic means (i.e., in TA's official website),[10] with safeguards of the taxpayer's personal data—article 68(17) of the GTL. Thanks to this rule, the interpretation of the law by the TA is publicly known, which brings greater legal certainty to taxpayers by assuring predictability in TA's future behavior.

As we will see below, there are already some published binding rulings by the TA regarding several situations related to cryptocurrencies, which allows economic actors to predict up to a certain degree the tax consequences of their acts (insofar as for cases similar to those in the binding rulings).

Additionally, the principle of good faith and trust is safeguarded in article 104(1) of the GTL. This legal provision establishes that the TA can be sentenced to a financial penalty under the rules on litigation in bad faith if it acts in court against the content of binding information previously provided to the taxpayer or when its behavior in the process differs from that usually adopted in identical situations. However, there are

---

[7] Gomes 1993, p. 145. The above-mentioned article 11(4) of the GTL is a guardian of the principle of legal certainty.

[8] Pires and Calçada Pires 2012, pp. 123–124. It is also from this principle that the prohibition on retroactivity of the tax law, provided by Article 103(3) of the Constitution, arises.

[9] This proposal is only mandatory if the request is urgent, case in which the ruling should be issued within 75 days under penalty of the proposal being tacitly sanctioned. Besides, a fee is due (the amount varies with the complexity). V. article 68(2), (7), (8) e (10) of the GTL.

[10] The binding rulings are published (only in Portuguese) at https://info.portaldasfinancas.gov.pt/pt/informacao_fiscal/informacoes_vinculativas/Pages/default.aspx.

practical limitations on binding rulings: the TA, pursuant the issuance of a binding ruling is not obliged to comply with all situations that fall within the scope of that same guideline. On the contrary, the connection of the TA to the content of the same is an *inter partes* connection since the TA is forbidden to proceed in a direction different from the ruling only in relation to the specific case of the requesting taxpayer.[11]

In short, despite the practical relevance of the existing binding rulings, the TA is only bound before the individual taxpayer who made the request and only regarding the facts covered by said request (in fact, a taxpayer may appeal against the ruling under article 68(29) of the GTL). Therefore, one needs to be careful when reading and interpreting a binding ruling, despite the "reprehensibility of changes in behavior" by the TA, taking into account the principles of good faith, trust and also equality.

Finally, if the TA gives an explanation in person, by telephone or by "*e-balcão*"[12] regarding a specific case (therefore, not within the scope of article 68 of the GTL), that answer is not binding, and caution is also desirable. Illustrating this point with a practical example, news on an "*e-balcão*" reply a query concerning PIT can be found online concluding that Bitcoin trading would give rise to taxable capital income, as it should be qualified as a profit distribution by a non-resident entity.[13] This interpretation, as will be seen, was thwarted by a binding ruling.

4.3 Taxing Crypto-assets under the Corporate Income Tax Code

4.3.1 The Corporate Income Tax Code: A Brief Introduction[14]

The Portuguese CIT is levied on all income obtained by the respective taxable persons (article 1 of the CIT Code). Article 2 of the CIT Code qualifies as taxable person several entities, such as (i) the commercial companies or civil companies in commercial form, cooperatives, public companies and other legal persons governed by public

---

[11] There is a different regime for generic information given by the TA on the interpretation of the tax rules that are in force at the time of the tax event, for which, under the terms of article 68-A(1) of the GTL, the TA is bound to comply with it in all situations that arise within the scope of that same orientation. It should be noted, however, that the administrative circulars issued by the TA are binding only for their services.

It should also be noted that there is no general information on tax issues (pursuant to article 68-A(1) of the GTL) related to crypto-assets, at least no information that is publicly available.

[12] "*E-balcão*" is a tool available at the TA official website via which one may submit questions and request information.

[13] Tocha 2019.

[14] At *guiadoinvestidor.dre.pt*. There are some small guides on the Portuguese tax system for investors available in English, including on the Corporate Income Tax.

or private law, with head office or effective management in Portuguese territory; (ii) entities without legal personality, with head office or effective management in Portuguese territory, whose income is not taxable on PIT or on CIT, respectively in the ownership of natural or legal persons;[15] and (iii) entities, with or without legal personality, that do not have head office or effective management in Portuguese territory and whose income obtained therein is not subject to PIT.

The taxable basis depends on the type of taxable person, under article 3 of the CIT Code, but for entities above under (i) that are primarily engaged in a commercial, industrial or agricultural activity[16] CIT is levied on profit (which consists of the difference between the net asset values at the end and the beginning of the tax period, with the corrections established in that Code).

Article 4 of the same Code establishes that for legal persons and other entities with head office or effective management in Portuguese territory, the CIT is levied on the totality of their income, including that obtained outside that territory (worldwide income principle) while for those without head office nor effective management in Portuguese territory the CIT is only levied on income obtained therein (territorial or source principle). The general CIT tax rate is 21% and there is a reduced tax rate of 17% to Small and Medium-Sized Enterprises for a taxable basis until € 25,000 (v. article 87(1) and (2) of the CIT Code).

4.3.2 Profits with Crypto-assets and the Importance of the Accounting Perspective

A taxable person under the CIT Code may obtain several sorts of income with crypto-assets that may contribute to its taxable profit. For instance, it may obtain gains from sales of crypto-assets, obtain commissions for rendering services related with these assets, obtain income derived from the sale of products or services that cover the crypto-asset industry or it may mine them. In all these cases, the taxation under the CIT Code is exactly the same that it would be if the taxable person obtained income with producing and selling chairs, render image consulting services or any other business activity since those activities are economic operations of a business nature and then perfectly covered by the CIT Code, namely its article 3.

---

[15] E.g., estates, invalid legal persons, associations and civil companies without legal personality and commercial companies or civil companies in commercial form, prior to definitive registration.

[16] Defined by article 3(4) of the CIT Code as all activities consisting of carrying out business-related economic operations, including the provision of services.

The TA has not yet issued any binding ruling, generic information or other interpretation regarding crypto-assets and CIT. However, the Portuguese Chartered Accountants Association ("OCC") published its Technical Opinion no. PT22246, of 22-03-2019 on cryptocurrencies.[17] This Technical Opinion, aside from explaining the binding rulings on PIT and VAT discussed below, has only one sentence on CIT: *"On CIT, there is no published position on the part of the Tax Authority, but we are of the opinion that the accounting income recorded in accordance with accounting standards contributes to taxation at CIT"*.[18]

Since CIT taxes the profit with the corrections established in its Code and the starting point is the accounting, it has a remarkable importance in this context, and one should ask (from an accounting perspective): are cryptocurrencies *stricto sensu* intangible assets or inventories? Are security tokens financial instruments? Are utility tokens intangible assets or contracts with clients?

Regarding holdings of cryptocurrencies, the International Financial Reporting Interpretations Committee ("IFRIC") released in June 2019 an update in which it discussed the nature of a cryptocurrency under the International Financial Reporting Standards ("IFRS"), concluding that it is an intangible asset under the International Accounting Standard ("IAS") 38 *"on the grounds that (a) it is capable of being separated from the holder and sold or transferred individually; and (b) it does not give the holder a right to receive a fixed or determinable number of units of currency"*, excluding the qualification as a financial asset (Paragraph 11 of IAS 32) or as Cash (Paragraph AG3 of IAS 32). Additionally, if an entity holds cryptocurrencies for sale in the ordinary course of business, that holding is Inventory (IAS 2).[19]

Furthermore, the Organization for Economic Cooperation and Development (OECD) stated that although there is currently no formal accounting guidance available on crypto-assets, it is necessary to apply the existing general accounting principles, namely their economic purpose, the rights and liabilities associated with them, and the way the assets derive their inherent value. Consequently, a security token could be considered as a financial asset subject to IFRS 9, and utility tokens can be considered as a prepayment for those goods and services, treated under IFRS 15.[20]

To conclude, although there are some uncertainties under the accounting perspective (which have impact in the CIT taxation), if a taxable person obtains income from a crypto-asset, such income will contribute for the taxable basis under the CIT Code, provided that the subjective and objective requirements explained above are fulfilled.

---

[17] Ordem dos Contabilistas Certificados [Portuguese Chartered Accountants Association] 2019.
[18] Free translation.
[19] International Financial Reporting Interpretations Committee 2019.
[20] OECD 2020, p. 15.

4.4 Taxing Crypto-assets under the Personal Income Tax Code

4.4.1 The Personal Income Tax Code: A Brief Introduction[21]

All natural persons resident in the Portuguese territory[22] are subject to PIT on all their income, including that obtained outside Portuguese territory (worldwide income principle) while non-resident natural persons are only taxed on income obtained therein (territorial or source principle)—article 15 of the PIT Code. For these taxpayers, we must analyze the taxable basis, as defined by article 1 of the PIT Code, which establishes the following:

> Article 1
>
> Taxable Basis
>
> 1—The personal income tax is levied on the annual value of the income from the following schedules, even when resulting from unlawful acts, after making the corresponding deductions and allowances:
>
> Schedule A—Employment income;
>
> Schedule B—Business and professional income;
>
> Schedule E—Capital Income;
>
> Schedule F—Property income;
>
> Schedule G—Capital gains;
>
> Schedule H—Pensions.
>
> 2—The income, whether in money or in kind, is subject to taxation, wherever it is obtained, the currency and the form in which it is earned.

Only the income from those schedules is taxed and the annual taxable amount is subject to progressive taxation up to 48%, plus an additional solidarity tax with a rate of 2,5% applicable to taxable income exceeding € 80,000 up to € 250,000 and of 5% applicable to taxable income exceeding € 250,000. However, there are some types of income taxed with autonomous or final tax rates—both usually of 28%.[23]

Therefore, in the case of natural persons, income from crypto-assets is taxed under the PIT Code only if it fits into one of these six schedules of income. And in that case, as it results from article 1(2), the taxation will be levied regardless the payment is made in kind or in cash (regardless of the currency used). This was the interpretative

---

[21] At *guiadoinvestidor.dre.pt*. There is also a small guide available on the Portuguese Personal Income Tax Code.

[22] Article 16 of the PIT Code defines who is resident in Portuguese territory. There are several special rules but the general one is that one is resident when one has been in Portugal over 183 (consecutive or interpolated) days in any 12-month period beginning or ending in the year in question or, if someone has stayed for a shorter time, has a dwelling there (in any day of the period referred to previously) in conditions that suppose someone's current intention to maintain and occupy it as a habitual residence.

[23] See articles 68, 68-A, 71 and 72 of the PIT Code.

path that the TA followed in the binding ruling described below in the following subsections.

4.4.2 The Sale of Cryptocurrencies

The first binding ruling in Portugal regarding crypto-assets was issued in 2016,[24] in which the applicant questioned what the legal framework for income was resulting from the purchase and sale of cryptocurrencies.

The TA started by asserting that cryptocurrencies "are not currencies" because they do not have legal tender or discharging power in Portugal. Further, the case under analysis was about gains obtained with the purchase and sale of virtual currency units or exchange at the exchange rate of cryptocurrency by fiat currency (whatever it may be). The income generated by this activity could, in theory, be integrated into three different schedules of income: G (capital gains), E (capital income) or B (business or professional income).

4.4.3 Capital Gains?

The TA analyzed whether this income could be considered capital gains under article 10 of the PIT Code (noting that it entails a closed typification), putting the following hypotheses:

- The TA considered that this is not a disposal of shares neither the cryptocurrencies confer rights to receive any amount;
- It is also not a derivative financial product, as its valuation is not based on any underlying asset;
- Finally, it defended that cryptocurrencies are not yet securities since they cannot be subsumed at this time to article 1 of the Securities Code.[25]

However, one may question if some crypto-assets could be qualified as atypical securities since article 1(g) of the Securities Code provides that *"Other titles representing similar legal situations provided they are market tradeable"* are considered securities. In fact, the Portuguese Securities Market Commission ("CMVM") issued a Press Release stating that a token launched in an Initial Coin Offering ("ICO") could be qualified as an atypical security if it is a document representing homogeneous legal situations susceptible to transmission on the market, but a case-by-case

---

[24] Portuguese Tax Authority 2016.
[25] Available in English at https://www.cmvm.pt/PInstitucional/Content?Input=E185223E6A3F59C DFEF430AA0FF3A98E6D788CDAAF3A590A381043B359120863.

Table 4.1 Example of USA

Cryptocurrencies qualification	Entity
Money	• Financial crimes enforcement network; Office of foreign assets control
Property	• Internal revenue services
Commodities	• Commodity futures trading commission
Securities	• Securities and exchange commission

Source The author[27]

analysis is always necessary, given the complexity and variability of this reality.[26] Therefore, the gains resulting from the purchase and sale of crypto-assets that can be qualified as securities would be capital gains taxable at an autonomous tax rate of 28%.

It is not unprecedented to have difficulties or differences in qualifying cryptocurrencies, namely in what concerns income resulting from their purchase and sale among official entities. On this see Table 4.1 below for the US example on qualification of cryptocurrencies by different relevant public authorities.

However, this can raise a problem in Portugal, since article 11(2) of the GTL establishes that whenever the tax rules use terms specific to other branches of law, they must be interpreted in the same manner as they are under such branches unless a different interpretation is directly provided by law. Thus, *prima facie* the TA should follow the interpretation made by CMVM. On the other hand, the TA interpretation gives more legal certainty because there is no need of casuistic analysis. Still, in light of CMVM communication (which was issued almost two years later than the binding ruling) and considering that the wording used by the TA was "at this time", it is not impossible that the TA changes its position. However, as far as we know, there is no indication that it is going to happen.

4.4.4 Capital Income?

Article 5(1) of the PIT Code establishes that income from capital are the legal fruits or any economic advantages, whatever their nature or denomination, whether in cash or in kind, arising, directly or indirectly, from patrimonial elements, assets, rights or legal situations, of a movable nature, as well as the respective modification, transfer or termination, with the exception of gains and other income taxed in other categories. The TA noted that this rule is drafted in an open manner and that article 5 provides

---

[26] Comissão do Mercado de Valores Mobiliários [Portuguese Securities Market Commission] 2018b. Indeed, in another Press Release CMVM analyzed the legal nature of Bityond cryptocurrency and concluded that it was not a security—Comissão do Mercado de Valores Mobiliários [Portuguese Securities Market Commission] 2018a.

[27] See, e.g., Ozelli 2018.

some examples of taxable events (but not the only ones). Thus, in schedule E the taxable income is the one resulting from the mere application of capital is taxed, i.e., the rights produced without impairment to the substance of the producer. Since the case under analysis was the income resulted from the sale of the right, the binding ruling concluded that it was not capital income.

4.4.5 Business or Professional Income?

Lastly, the binding ruling analyzed the case under schedule B, enunciating that in such case the income is taxed according to the exercise of an activity and not according to the origin of the income (i.e., regardless of whether the income results from sales, legal fruits, or has any other nature), under article 3(1) of the PIT Code. Further, the exercise of the activity is determined by its regularity and by the orientation of such activity to obtain profits.

In case of existence of a business or professional activity, the taxpayer is obliged to comply with the declarative obligations contained in article 3(6) of the PIT Code—to issue an invoice or equivalent document (electronic invoice-receipt), whenever he or she makes a sale or provides a service. In this case, taxation occurs at general progressive rates.

Then the binding ruling concluded that the sale of crypto-currency is not taxable under the Portuguese tax system unless if by its regularity it constitutes a professional or business activity of the taxpayer, in which case it will be taxed under schedule B. Therefore, one should question how to differentiate a regular activity from an occasional activity? These are indeterminate concepts and disputes may arise from their interpretation. However, we are not aware of any cases in which the TA is taxing this income, apart from those in which the taxpayer declared the beginning of a business and professional activity related with cryptocurrencies. In fact, the ECJ (albeit with regard to VAT), has already stated that the practice of a certain set of acts of mere exercise of wealth management is not, *per se*, likely to be seen as the result of the exercise of an economic activity.[28]

---

[28] See e.g., ECJ, Wellcome Trust, Judgment, 20 June 1996, C-155/94 and ECJ, Enkler, Judgment, 26 September 1996, C-230/94.

If a taxpayer does not declare himself/herself to be pursuing a professional or business activity, the burden of proof will be on TA, which leads to compliance issues. The European Commission has an initiative in progress to amend the Directive on Administrative Cooperation ("DAC") to ensure that EU rules stay in line with the evolving economy and include other areas such as crypto-assets and e-money, looking for progress on tax transparency and deal with the substantial risks of tax evasion. The adoption of a proposal for a DAC8 is planned for third quarter of 2021. The initiative is available at https://ec.europa.eu/info/law/better-regulation/have-your-say/initiatives/12632-Tax-fraud-&-evasion-strengthening-rules-on-administrative-cooperation-and-expanding-the-exchange-of-information_en Accessed 30 May 2021.

4.4.6 Other Types of Income and Some Practical Considerations

There are no other public understandings of the TA regarding PIT and crypto-assets. Nevertheless, here follow some additional thoughts, considering that article 1(2) of the PIT Code determines the taxation of income, whether paid in cash or in kind. In our opinion, crypto-assets may be qualified as payments in kind and if the parties under a contract agree with such a form of payment, there is no obstacle to that from a tax perspective. Therefore, for instance, if one receives crypto-assets as payment for goods sold or services rendered under schedule B, it will be taxed at the general progressive tax rates. Or if a salary is paid with cryptocurrencies, it will be taxed under schedule A, also at the general progressive tax rates. If a landlord accepts crypto-assets to settle the rent, it will be taxed under schedule F with an autonomous taxation of 28%.

Still, one practical difficulty may arise: what is the exchange rate to Euros that one should apply? The PIT Code establishes in its article 23 the rules for values fixed in a currency without legal tender in Portugal (the equivalence to Euro should be determined by the official quotation of the respective currency, at the day resulting from those rules). However, this is not the appropriate solution since crypto-currencies are not technically currency.[29] Instead, we should apply article 24 of the same Code, which provides successive rules for cash equivalence for income in kind:

(a) At the officially established price;
(b) By the official purchase quotation;
(c) In the case of goods, by the purchase quotation on the Lisbon merchandise exchange or, if there is no such quotation, by the average price of the respective year or the last one determined, and which are included in the city stowage;
(d) For the prices of similar goods or services published by the National Statistics Institute;
(e) At market value, under conditions of competition.

Since there is no official prices or quotations for crypto-assets, in our opinion, we should use market value (available in exchanges platforms). This was, for instance, the IRS solution on income resulting from an airdrop of crypto-currency in the sequence of a hard fork, determining that the taxable basis is measured by the respective market value.[30]

---

[29] According to the TA, as we saw above, and also ECJ as we will see below.
[30] Internal Revenue Service 2019.

4.5 Taxing Crypto-assets under the Value Added Tax Code

4.5.1 The Value Added Tax Code: A Brief Introduction[31]

The transfers of goods and the provision of services (as defined, respectively, in articles 3 and 4 of the VAT Code) carried out in Portugal, for consideration, by a taxable person acting as such, imports of goods (article 5 of the same Code) and intra-Community operations carried out in the national territory, as defined and regulated in the VAT Regime for Intra-Community Transactions are subject to VAT—v. article 1 of the VAT Code.

Article 2 of the VAT Code establishes who are the taxable persons, among which we should highlight the natural or legal persons who, in an independent and habitual manner, carry out activities of production, trade or provision of services, including extractive, agricultural and free-trade activities, as well as those who, of the same independently carry out a single taxable transaction, provided that such transaction is connected with the exercise of the aforementioned activities, wherever this occurs, or when, regardless of that connection, such transaction fulfils the real impact assumptions of the PIT or CIT (paragraph 1(a) of the same article).

There are some exemptions, namely the ones related to the types of activity (article 9 of the VAT Code) and those related to turnover (article 53 of the same Code). The VAT rates are defined in article 18 of the Code and in the attached tables, being the general tax rate 23%, the intermediate tax rate 13% and the reduced tax rate 6% (for mainland Portugal).

4.5.2 The European Court of Justice Hedqvist Case[32]

In this landmark case there was a request for a preliminary ruling on whether transactions to exchange a traditional currency for the Bitcoin or vice versa (held by Mr. Hedqvist's company) were subject to VAT. Article 2 of the VAT Directive[33] establishes that supply of goods and supply of services (with some territoriality rules) are transactions subject to VAT. Under article 14(1) of the same Directive, "supply of goods" shall mean the transfer of the right to dispose of tangible property as owner, while under article 24(1) "supply of services" shall mean any transaction which does not constitute a supply of goods.

The ECJ started by noting that virtual currencies with bidirectional flow are not electronic money[34] according to a 2012 report by the European Central Bank because

---

[31] At *guiadoinvestidor.dre.pt*—it is also possible to find a small guide on the Portuguese VAT Code.
[32] ECJ, Hedqvist, Judgment, 22 October 2015, C-264/14.
[33] Council Directive 2006/112/EC of 28 November 2006 (as amended).
[34] As defined in Directive 2009/110/EC of the European Parliament and the Council of 16 September 2009.

unlike that specific type of funds, virtual currencies are not expressed in traditional accounting units (e.g. Euro), but in virtual accounting units (e.g. Bitcoin). Those virtual currencies are not also "tangible property" (under article 14 of the VAT Directive) because its only purpose is to be a mean of payment. Since the type of exchange in question consisted in an exchange of different means of payment, the ECJ decided that it cannot be qualified as a "supply of good" and consequently it is instead a "supply of services".

Additionally, the ECJ decided that there is a consideration[35] in this supply of services because there was direct link between the services provided and the consideration received by the taxable person, being irrelevant that this remuneration does not take the form of payment of a commission or the payment for specific expenses (quoting the Case First National Bank of Chicago).[36]

Subsequently, the ECJ concluded that article 135(1)(d) of the VAT Directive (which exempts "*transactions, including negotiation, concerning deposit and current accounts, payments, transfers, debts, cheques and other negotiable instruments, but excluding debt collection*") was not applicable because Bitcoin is a contractual mean of payment. However, it considered that the exemption provided by article 135(1)(e) ("*transactions, including negotiation, concerning currency, bank notes and coins used as legal tender, with the exception of collectors' items, that is to say, gold, silver or other metal coins or bank notes which are not normally used as legal tender or coins of numismatic interest*") was applicable because if it covered only traditional currencies, the ECJ would be depriving such provision of a part of its effects since it is intended to alleviate the difficulties associated with determining the taxable amount and the amount of deductible VAT. Consequently, the transactions *sub judice* are exempted from VAT since they were covered by article 135(1)(e) of the VAT Directive.

Lastly, the ECJ stated that this type of transactions do not fall under the scope of article 135(1)(f)of the same Directive ("*transactions, including negotiation but not management or safekeeping, in shares, interests in companies or associations, debentures and other securities, but excluding documents establishing title to goods, and the rights or securities referred to in Article 15(2)*") because Bitcoin was neither a security conferring a property right nor a security of a comparable nature.

4.5.3 The Sale of Cryptocurrencies

There is also a binding ruling in Portugal regarding the sale of cryptocurrencies for VAT purposes in which the applicant questioned what the legal framework is (and

---

[35] Article 2(1) of the VAT Directive provides that the supply of goods and services for consideration within the territory of a Member State by a taxable person acting as such is to be subject to VAT.

[36] ECJ, First National Bank of Chicago, Judgment, 14 July 1998, C-172/96.

also which are the documental procedures) for sales of Bitcoin through an electronic platform.[37]

The TA began by clarifying that Bitcoin is a virtual currency and that the transaction in question would be a costly service provision (art. 4 of the VAT Code) because it is not a transfer of tangible goods. If these transactions are carried out against consideration by a VAT taxable person, in the exercise of an economic activity, they constitute transactions subject to tax (articles 1 (1) (a) and 2 (1) (a) of the VAT Code), where the taxable amount corresponds to the amount of the consideration (article 16 (1) of the same Code).[38]

However, the transaction will be exempt from VAT because Article 9(27)(d) of the VAT Code exempts *"transactions, including trading, whose object is currency, bank notes and coins, which are legal means of payment, with the exception of coins and banknotes that are not normally used as such, or that have a numismatic interest"*.

Therefore, this binding ruling closely follows the Hedqvist Case and shows that it is necessary that a VAT taxable person pursues an economic activity (under schedule B of PIT or CIT), i.e., that the activity is carried out in an independent and habitual manner (art. 2 of VAT Code).

4.5.4 Mining

The last binding ruling on these matters published by the TA was on remuneration with cryptocurrency through mining.[39] The TA noted that mining qualifies as a blockchain transaction organization paid with cryptocurrency, but the activity is not provided to anyone in particular. Based on the Hedqvist Case, the TA concluded that the exchange of cryptocurrency for "real" currency (i.e., fiat currency) is an onerous service, subject but exempt from VAT (article 9(27)(d) of the VAT Code), and so it is the remuneration in cryptocurrency, given that it is a contractual means of payment. Further, regarding the location of operations, and since the way the question was made does not allow to individualize the operations, there was a remission to generic clarifications provided in some Circular Letters.

However, despite this ruling, it is doubtful that there is an actual consideration, not only because in the case above the activity is not provided to anyone in particular (instead, it is provided to a "pool"[40] and not to a managing entity for instance, within a legal relationship), but mostly because it is difficult to find a direct link if not

---

[37] Autoridade Tributária e Aduaneira [Portuguese Tax Authority] 2019a.

[38] Again, there may be practical difficulties with the exchange rate, as art. 16 (8) of the VAT Code states that *"When the elements necessary to determine the taxable amount are expressed in a currency other than the national currency, the exchange rate to be used is the last one disclosed by the European Central Bank or the sale rate practiced by any bank established in the national territory."*.

[39] Autoridade Tributária e Aduaneira [Portuguese Tax Authority] 2019b.

[40] Arguing the in mining there is no reciprocal performance because there is no specific customer—Wolf 2014, p. 257.

Table 4.2 Value added tax committee's conclusions on mining

"Activity	Subject to VAT?	If so, exempt?
Mining Activities	• *Out of scope*: The fact that the payment of a transaction fee by a Bitcoin user is not a necessary condition for successfully sending bitcoins (and thus for receiving a verification service supplied by the miner) may be indicative of there not being a direct link between the consideration and the service. Besides, the provision of a mining service does not create for the miner the right to receive a consideration in exchange, which could imply the nonexistence of a legal synallagmatic relationship between him and the recipient of the verification services (the user whose transaction request the miner has validated).	
	• *Taxable:* New bitcoins received automatically by the miner from the Bitcoin system every time that a verification service is supplied could possibly be seen as constituting a consideration for a taxable service. • Despite the fact that Bitcoin transactions carried out for free are in theory possible, in practice Bitcoin users pay fees (used as a default by most digital wallets); and it seems almost impossible to imagine users would be willing to wait days or weeks before a transaction is verified (which could be the case if no fee is paid).	• *Exempt:* Mining activities could be seen as exempt pursuant to Article 135(1)(e) of the VAT Directive, on the grounds of them being services directly concerning currency. • *Exempt:* Mining activities could be treated as exempt pursuant to Article 135(1)(d) of the VAT Directive on the basis of them fulfilling in effect the specific, essential functions of an exempt supply (the transfer of bitcoins itself)."

Source Value Added Tax Committee—European Commission 2016

all mining activity leads to the "production" of a cryptocurrency.[41] Additionally, the Hedqvist decision does not comment on mining. So, even though the TA bases its interpretation on such decision, it does not mean that it is necessarily the correct position. In fact, the VAT Committee issued a working paper on several issues arising from such judgment, concluding that (Table 4.2).

Lastly, one should note that precisely because the ECJ's decision does not cover mining, and as Table 4.3 below exemplifies, there is still no VAT neutrality in the EU:

---

[41] In the same sense that there is no guarantee that miners will receive cryptocurrency, therefore putting in question the onerous nature, see Gamito et al. 2015, p. 119.

Table 4.3 Mining VAT Treatment in Some EU Member States

State	Taxation
France	Revenue received from cryptocurrency mining activities is subject to VAT as a supply of services.
Germany	Revenue received from cryptocurrency mining activities is generally outside the scope of VAT.
Spain	Revenue received from cryptocurrency mining activities is generally outside the scope of VAT.
Sweden	Revenue received from crypto mining activities is generally outside the scope of VAT.

Source The author[42]

4.6 Conclusions

In the crypto-assets context, one should note the importance of the principle of tax legality because if the facts do not fulfil the provision of a statutory tax law, there is no taxable event (so the taxpayer should not be subject to tax). The principle of legal certainty is also relevant, especially in new and complex areas. Binding rulings by the Portuguese TA are an important available instrument to any taxpayer since they clarify the interpretation of tax law. They are publicly known, which brings greater legal certainty to taxpayers, through the predictability of the TA's behavior. However, the TA is only bound on those facts and to the taxpayer who submitted the request.

A taxable person under the CIT Code may obtain several sorts of income with crypto-assets and all of them contribute for its taxable profit since those activities are economic operations of a business nature and then perfectly covered by the CIT Code, namely its article 3. As the OCC noted, the accounting income should consider the accounting standards, namely the IFRIC updates.

In the case of natural persons, income from crypto-assets is taxed under the PIT Code only if we can fit it into one of the six schedules of income. The TA concluded that income resultant from the sale of cryptocurrencies is not capital gains because they are not securities. Still, one may question whether some crypto-assets are actually atypical securities under article 1(g) of the Securities Code. TA's interpretation gives more legal certainty because there is no need of casuistic analysis. Still, in light of the CMVM communication and considering that the wording used by the TA was "at this time", it is not impossible that the TA changes its position. However, there is no indication that it is going to happen. Instead, the TA concluded that the sale of crypto-currency is not taxable under the Portuguese tax system unless by its regularity it constitutes a professional or business activity of the taxpayer, in which case it will be taxed as business or professional income. Additionally, considering that the income of any schedule can be paid in cash or in kind, if a PIT income is paid with crypto-assets, we should apply article 24 of the same Code, which provides successive rules for cash equivalence for income in kind.

---

[42] Based on Jupe 2018.

Lastly, in the Hedqvist case, ECJ concluded the sale of bitcoins was a supply of services VAT exempt by article 135(1)(e) of the VAT Directive, although they are not money since virtual and traditional currency should have the same treatment for the purposes of that exemption. Following this case, the Portuguese TA considered that the sale of cryptocurrencies was also exempted VAT under Article 9(27)(d) of the VAT Code. Finally, also based in the same case, the Portuguese TA concluded that mining would benefit from the same exemption. However, as shown by the VAT Committee (and also the position of other Member States), from that decision it is possible to conclude that mining is out of VAT scope, taxable, exempt under article 135(1)(e) or even exempt under article 135(1)(d). Therefore, even in VAT there are also some uncertainties on some activities related to crypto-assets.

4.7 Afterword[43]

Since the last revision of the chapter, several relevant developments occurred that we believe important to be highlighted. The first is the significant impact that the **Portuguese Budget State Law for 2023**[44] had on the taxation of crypto-assets. The legislator, taking into consideration the principles of tax legality and legal certainty in the context of crypto-assets, decided to create specific tax rules for crypto-assets.[45]

In this line, the PIT Code has now a **definition of crypto-assets**: "*any digital representation of value or rights that may be transferred or stored electronically using distributed ledger or similar technology*". This concept does not include unique crypto-assets that are not fungible with other crypto assets, i.e., Non-Fungible Tokens (NFT).[46] The legislator also clarified that article 24 of the PIT Code applies, i.e., the cash equivalence of crypto-assets follows the same rules already established for income in kind (v. Sect. 4.4 *supra*).

Another relevant amendment to PIT Code is the new regime in case of **sale of cryptocurrencies and other crypto-assets**: article 10(1)(k) of PIT Code now qualifies the onerous transfer of crypto-assets (that are not securities)[47] in schedule G (capital gains), provided that it occurs outside the scope of a business activity or professional (case in which Schedule B applies). The capital gain is thus subject to a tax rate of 28%, with the option for the general tax rates. However, gains and losses arising from the sale of crypto-assets (that are not securities) held for at least 365

---

[43] Written: 25 April 2023.

[44] Law no. 24-D/2022, December 30.

[45] The Report of the Proposal of the Budget State Law for 2023, parlamento.pt/statebudget2023, accessed 25 April 2023, pp. 74–75, stated that the intention was to create a broad and adequate tax framework applicable to crypto-assets, in terms of income and property taxation, providing security and legal certainty by creating a specific regime that aims to promote the crypto-economy and projecting the digital transition and expanding the economy.

[46] Article 10(17) and (18) of the PIT Code. This definition is also applicable for CIT and Stamp Tax purposes.

[47] Seeming to confirm my previous interpretation that crypto-assets that are securities would already be taxed—v. *supra* Sect. 4.4.

days are excluded from taxation.[48] In addition, income is only subject to tax when the consideration is not crypto-assets (article 10(19) and 20 of the PIT Code).[49] An exit tax was also included, according to which the loss of the status of tax resident in Portugal is equivalent to a taxable onerous transfer (article 10(22) of the PIT Code).

As for other types of income, it is relevant to note the regime of **schedule B (business and professional income)**: in article 4(1)(o) of the PIT Code, there is now an express qualification as commercial and industrial activity of operations related to the issuance of crypto-assets, including mining, and with the validation of crypto-asset transactions through mechanisms of consensus. Furthermore, within the scope of the simplified taxation regime,[50] it is now expressly foreseen the calculation of the taxable income through the application of the coefficient of 0.15 to operations with crypto-assets, with the exception of mining of crypto-assets, to which the coefficient of 0.95 is applied[51] (article 31(1)(a) and (d) of the PIT Code).

It should also be noted that now article 5(2)(u) of the PIT Code, qualifies as **capital income (schedule E)** any forms of remuneration resulting from operations related to crypto-assets (subject to a rate of 28%, with an option for taxation at general tax rates). This income is exempt from withholding at source (regardless of the form it takes—that is, even when paid in fiat) and, when it is paid in the form of crypto-assets, it is taxed under category G (capital gains) in accordance with article 5(11), in which case the above rules apply.

As for **Corporate Income Tax**, what was written in chapter III remains unchanged. However, in CIT there is also a simplified regime, that now provides for the inclusion of income related to crypto-assets in determining the taxable amount, applying the coefficients 0.95 in the case of income from mining crypto-assets and 0.15 in the case of other income related to crypto-assets (article 86.-B (1)(b) and (e) of CIT Code). As for the **Value Added Tax (VAT)**, what was said in Chapter V has not changed.[52]

---

[48] Article 220 of the State Budget Law included a transitory rule according to which the holding period prior to the entry into force of the law should also be considered.

[49] However, this exclusion does not apply to income earned by taxable persons or owed by any person/entity that are not tax residents in another EU or EEA MS or in another State where there is a Convention to Avoid Double Taxation or a bilateral or multilateral agreement that provides for the exchange of information for tax purposes—article 10(21) of the PIT Code.

[50] Under this regime (article 31 of the PIT Code), taxable income is obtained through the application of coefficients, in which only the result of the application of that rate will be taxed.

[51] As recommended by the European Commission in its Communication "Digitalizing the energy sector—EU action plan"—https://ec.europa.eu/info/law/better-regulation/have-your-say/initiatives/13141-Digitalising-the-energy-sector-EU-action-plan_en, accessed 25 April 2023.

[52] However, and despite not being specifically for the Portuguese context as a result of the 2023 State Budget, it should be noted that the VAT Committee recently issued a working paper qualifying the issuance of NFTs in general as an electronic service but noting that a case-by-case analysis is necessary, since substance should prevail over form (considering the principle of VAT neutrality). V. Value Added Tax Committee—European Commission (2023), Working Paper No. 1060, "Initial VAT reflections on non-fungible tokens" https://circabc.europa.eu/rest/download/7d1ef2eb-b820-4866-a155-785e2373fb80?ticket=, accessed 25 April 2023. For an analysis of the same, see Costa Monteiro 2023.

A novelty is the express provision that in the case of acquisition of real estate with crypto-assets, i.e., exchange between real estate and crypto-assets, for the purpose of determining the taxable basis of **Municipal Tax on Onerous Transfers of Real Estate (IMT)**,[53] the value of the act or the contract (i.e., the "price") should be considered the value of the crypto-assets given in exchange, determined under the terms of the Stamp Tax Code (article 12(5(b) of the IMT Code).

Another important legislative novelty is in terms of **Stamp Tax**:[54] free transfers (donations and inheritances) of crypto-assets are now taxed at a rate of 10%,[55] when they are deposited in institutions in Portugal or, even if not being deposited, if the author of the transfer was domiciled in Portugal, in the case of inheritances, or if the beneficiary is domiciled in Portugal, in the case of donations (articles 1(3)(i) and 4(4)(e) of the Stamp Tax Code).[56] Thus, crypto-assets are now included in the list of values for which custody service providers, in the case of free transfers, can only allow the withdrawal after proof of Stamp Tax payment or declaration, if the transfer is exempt (article 63-A of the Stamp Tax Code).

Additionally, commissions and fees charged by or through the intermediation of crypto-assets service providers are also taxed at a rate of 4%, when the provider or client is domiciled in Portugal, being the tax charged to the client (item 30 of Table General of the Stamp Tax and articles 3(3)(w) and 4(9) of the Stamp Tax Code). It is determined that, when the service provider is located outside Portugal, the person who has to settle the tax is the intermediary in Portugal or, if there is no intermediary, a representative appointed by the service provider (article 2(1)(u) of Stamp Tax Code).

Finally, in order to enable not only compliance but also enforcement, article 124-A PIT Code also includes **reporting obligations**, by establishing that custody service providers and trading platform managers have to report annually all operations with crypto-assets performed with their intervention to the Tax Authorities. Also on reporting obligations (this will not amend the taxation of crypto-assets), a relevant international new standard results from an **Organization for Economic Cooperation and Development (OECD)** publication known as the **Crypto-Asset Reporting Framework (CARF)**.[57] This reporting framework intends to implement the automatic exchange of information regarding crypto-assets, comprising in its scope all

---

[53] On IMT in general, v. https://guiadoinvestidor.dre.pt, accessed 25 April 2023.

[54] On the Stamp Tax in general, v. https://guiadoinvestidor.dre.pt/Detail.aspx?TopicoId=22&AreaEnquadramento=2&OpcoesId=0&Lingua=pt-PT&PalavraChave=Impostodes, accessed 25 April 2023.

[55] However, there is an exemption when the free transfer occurs between spouses or de facto partners, descendants, and ascendants (article 6(1)(e) of the Stamp Tax Code).

[56] The Code includes in article 14-A rules on how to determine the taxable amount.

[57] OECD (2022) Crypto-Asset Reporting Framework and Amendments to the Common Reporting Standard, OECD, Paris, https://www.oecd.org/tax/exchange-of-tax-information/crypto-asset-reporting-framework-and-amendments-to-the-common-reporting-standard.htm, accessed 25 April 2023. The public consultation document was first released in March 2022 and the final guidance was published on 10 October 2022. It comprises solely reporting rules and will not amend national taxation regimes.

types of crypto-assets[58] (including NFTs) but excluding Central Bank Digital Currencies (CBDCs),[59] Specified Electronic Money Products and crypto-assets that cannot be used for payment or investment purposes. The reportable transactions will be (i) exchanges between crypto-assets and fiat; (ii) exchanges between one or more types of crypto-assets; (iii) transfer of crypto-assets in consideration of goods or services (with minimum thresholds) and (iv) transfers of crypto-assets. The reporting obligation will be due by all intermediaries that provide services regarding crypto-asset transactions for or on behalf of customers.

Lastly, on 8 December 2022, the European Commission adopted the already planned **proposal for a Council Directive amending Directive 2011/16/EU (DAC8 proposal)**[60] (v. *supra* footnote 28). In addition to other issues, DAC8 will extend the existing scope of automatic exchange of information under the Directives on administrative cooperation in the field of taxation (DAC), to cover information on crypto-assets and e-money transactions, that should be reported on an annual basis. However, the scope of DAC8 is broader than the OECD CARF standards, since the DAC8 rules would also apply to non-EU crypto-asset operators with reportable users in the EU. The reporting operators must transmit information regarding the type of asset, its transferor and its acquirer, and in some cases the reporting obligation will also cover digital tokens, more specifically, NFTs. The proposal is designed to be implemented in the EU Member States in 2026, this being the first year to present the reports enshrined in the DAC8.

References

Costa Monteiro P (2023) NFT's—VAT Committee's Initial Reflections, in Sharing Tax Thoughts, Nova Tax Research Lab. Available at https://taxlab.novalaw.unl.pt/?page_id=3545. Accessed 25 April 2023

Da Guerra Fonseca R, Otero P (2008) Comentário à constituição portuguesa, Volume II. Almedina Editora, Coimbra

---

[58] "The term "Crypto-Asset" is intended to cover any digital representation of value that relies on a cryptographically secured distributed ledger or a similar technology to validate and secure transactions, where the ownership of, or right to, such value can be traded or transferred to other individuals or Entities in a digital manner. As such, the term "Crypto-Asset" encompasses both fungible and non-fungible tokens and therefore includes non-fungible tokens (NFTs) representing rights to collectibles, games, works of art, physical property or financial documents that can be traded or transferred to other individuals or Entities in a digital manner."—p. 47.

[59] On this topic, see Chap. 7 by Beja and Correia Barradas in this book.

[60] Available at https://eur-lex.europa.eu/legal-content/EN/TXT/HTML/?uri=CELEX:52022PC0707&from=EN. The proposal took into consideration the Markets in Crypto-assets (MiCA) Regulation proposal and the OECD CARF. For an analysis of this proposal, v. Popa and Valério 2023.

Dourado AP (2007) O Princípio da Legalidade Fiscal—Tipicidade, conceitos jurídicos indeterminados e margem de livre apreciação. Almedina Editora, Coimbra

Dourado AP (2014) No Taxation without Representation in the European Union: Democracy, Patriotism and Taxes. In: Brokelind C (ed) Principles of Law: Function, Status and Impact in EU Tax Law. IBFD, Amsterdam. https://www.cideeff.pt/xms/files/Artigos_APD/Principles_of_law_chapter_10.pdf Accessed 5 May 2020

Gamito C, Antas F, Branco Pires J (2015) Bitcoins e IVA. Cadernos IVA. 2015:109–136

Jupe E (2018) Taxation of cryptocurrencies in Europe: an overview. https://www.osborneclarkefintech.com/2018/12/19/taxation-of-cryptocurrencies-in-europe-an-overview/ Accessed 29 September 2019

Ozelli S (2018) U.S. Tax Implications of Cross-Border Cryptocurrency Bribes. Tax Notes. 91:9. 895:901

Pires M, Calçada Pires R (2012) Direito Fiscal, 5th edn. Almedina Editora, Coimbra

Popa O, Valério C (2023) The (Most Recent) Proposal for an EU Directive to Amend the Rules on Administrative Cooperation in the Field of Taxation (DAC8), European Taxation, (Volume 63), No. 2/3

Sá Gomes N (1993) O princípio da segurança jurídica na criação e aplicação do tributo. Ciência e Técnica Fiscal. 371:141–178

Tocha C (2019), Bitcoin no IRS: será esta moeda virtual isenta de impostos? Ekonomista. https://www.e-konomista.pt/bitcoin-no-irs/ Accessed 6 September 2020

Wolf R (2014) Bitcoin and EU VAT. International VAT Monitor. September/October 2014. 254:257

Other Documents

Internal Revenue Service (2019). Revenue Ruling 2019-24. https://www.irs.gov/pub/irs-drop/rr-19-24.pdf Accessed 14 October 2019

International Financial Reporting Interpretations Committee (2019) Holdings of Cryptocurrencies—IFRIC Update June 2019. https://www.ifrs.org/news-and-events/updates/ifric-updates/june-2019/ Accessed 20 September 2019

OECD (2020) Taxing Virtual Currencies: An Overview Of Tax Treatments And Emerging Tax Policy Issues, OECD, Paris. www.oecd.org/tax/tax-policy/taxing-virtual-currencies-an-overview-of-tax-treatments-and-emerging-tax-policyissues.htm Accessed 21 March 2020

Ordem dos Contabilistas Certificados [Portuguese Chartered Accountants Association] (2019) Parecer técnico no [Technical Opinion no.] PT22246. https://www.occ.pt/pt/noticias/criptomoedas/ Accessed 10 September 2019

Portuguese Tax Authority (2016) Case no. 5717/2015, with an order dated of 27-12-2016, from the Deputy Director-General for Income Taxes. https://info.portaldasfinancas.gov.pt/pt/informacao_fiscal/informacoes_vinculativas/rendimento/cirs/Documents/PIV_09541.pdf Accessed 14 October 2019

Portuguese Tax Authority (2019a) Case no. 14763, with an order dated of 28-01-2019a, from the Director for VAT https://info.portaldasfinancas.gov.pt/pt/informacao_fiscal/informacoes_vinculativas/despesa/civa/Documents/INFORMACAO_14763.pdf Accessed 14 October 2019a

Portuguese Tax Authority (2019b) Case no. 14436, with an order dated of 2019b-07-03, from the Director for VAT. https://info.portaldasfinancas.gov.pt/pt/informacao_fiscal/informacoes_vinculativas/despesa/civa/Documents/INFORMACAO_14436.pdf Accessed 14 October 2019b

Portuguese Securities Market Commission (2018a) Press Release on Bityond cryptocurrency of 17 May 2018a. https://www.cmvm.pt/en/Comunicados/Comunicados/Pages/2018a0517mc.aspx Accessed 29 September 2019

Portuguese Securities Market Commission (2018b) Press Release to the entities involved in the launch of "Initial Coin Offerings" (ICOs) regarding the legal qualification of the tokens of

23 July 2018b. https://www.cmvm.pt/pt/Comunicados/Comunicados/Pages/2018b0723a.aspx Accessed 29 September 2019

Value Added Tax Committee—European Commission (2016). Working Paper No 892—CJEU Case C-264/14 Hedqvist: Bitcoin. https://circabc.europa.eu/sd/a/add54a49-9991-45ae-aac5-1e260b136c9e/892%20-%20CJEU%20Case%20C-264-14%20Hedqvist%20-%20Bitcoin.pdf Accessed 1 September 2020

Chapter 5
Blockchain Execution of Judgements—A Possibility in South Africa?

George Daniel Raath

Contents

5.1	Introduction	76
5.2	Traditional Methods of Execution of Judgments Under South African Law	76
5.3	Smart Contracts	79
5.4	Application in the South African Legal Context	81
	5.4.1 Execution	81
	5.4.2 Practical Impediments to Implementation of Blockchain Execution of Court Orders	82
	5.4.3 Benefits to the Parties	84
	5.4.4 Security for Costs	86
5.5	Conclusion	87
References		87

Abstract Executing court orders via conventional methods in South Africa can be costly and time consuming. An alternative to the *status quo* could be on-blockchain execution using smart contracts. This would afford the parties the dual benefits of irrevocability and immutability—thus, speed of execution and certainty of payment. This chapter considers conventional execution methods available to a successful litigant under South African law (Sect. 5.2). It then considers the concept and attributes of execution on the blockchain (Sect. 5.3), as well as two possible use cases of blockchain execution applied in a South African context and possible impediments to their adoption (Sect. 5.4).

Keywords Execution of judgments · Blockchain · Smart contracts · South African law

---

G. D. Raath (✉)
MinterEllison, Constitution Place, 1 Constitution Avenue, Canberra, Australia
e-mail: daniel.raath@minterellison.com

© T.M.C. ASSER PRESS and the author(s) 2024
F. Pereira Coutinho et al. (eds.), *Blockchain and the Law*, Information Technology and Law Series 37, https://doi.org/10.1007/978-94-6265-579-9_5

5.1 Introduction

As any litigant will know who has obtained a court order or arbitration award that remains unsatisfied post litigation, an order without the means to successfully execute it is largely meaningless.[1] Executing orders using blockchain technology—which affords the crucial dual benefits of irrevocability (and thus finality) and immutability—could provide a more efficient method of satisfying orders than is presently available in South Africa under conventional methods of execution.

This chapter will consider the possibility of using blockchain to execute court orders, from the vantage point of a litigant in South Africa, relying on South African law.

In doing so, the chapter will first consider conventional execution methods available to a successful litigant under South African law (Sect. 5.2). It will then consider the concept and attributes of execution on the blockchain (Sect. 5.3). It will thereafter consider two possible use cases of blockchain execution applied in a South African context and possible impediments to their adoption (Sect. 5.4).

5.2 Traditional Methods of Execution of Judgments Under South African Law

Consider the following scenario: Party A institutes a commercial claim in South Africa against Party B. Both companies are domiciled in South Africa. The claim is brought either in the High Court, via conventional litigation, or in arbitration proceedings pursuant to an arbitration agreement concluded between the parties. Party A, for reasons relevant to the suit, is required to furnish Party B with security for costs.[2] Party A is successful in its claim, obtaining an order for monetary judgment against Party B.

Party A, now holding a judgment against Party B, is entitled to execute on its judgment against the assets of party B.[3] As the execution takes place in South Africa and in terms of a South African High Court order, the traditional execution process is

---

[1] Harms 2016, para 621 explains the position thus: "A procedure whereby a successful litigant (the judgment creditor) can enforce the judgment is crucial to the legal process. The idea that a court must be able to give an effective judgment is the foundation of the law of jurisdiction. Effectiveness is inextricably linked with execution because, if the court through its officers has no control over either the defendant or the defendant's property, a judgment granted in favour of a plaintiff could amount to little more than a declaration of a theoretical benefit."

[2] The general rule under South African law is that peregrinus plaintiffs or applicants who are subject to the jurisdiction of the court are liable to provide security for the costs of litigation in which they are engaged. See Harms 2016, para 31.

[3] In this context, execution is meant in its technical legal sense of the process by which practical effect is given to the terms of a court order. See Harms 2016, para 621.

governed by Rules 45 and 46 of the Uniform Rules of Court (**the Uniform Rules**).[4] The process is lengthy, arduous and expensive.

As a starting point, Rule 45 requires the judgment creditor (Party A) to have issued with the registrar of the High Court a writ of execution for attachment of the assets of the judgment debtor (Party B).[5]

This having been done, the writ is provided to the sheriff of the court, who is required to attend at the judgment debtor's place of business to *"demand satisfaction of the writ and, failing satisfaction...demand that so much movable and disposable property be pointed out as he may deem sufficient to satisfy the said writ, and failing such pointing out...search for such property."*[6]

Once the sheriff has prepared an inventory of the goods attached, they are sold at an execution sale, by public auction to the highest bidder, to satisfy the judgment in whole or in part.[7]

Rule 45 also makes provision for the attachment by sheriff of incorporeal assets (such as shares or claims) pursuant to a writ issued for the attachment of those assets,[8] or the attachment of income due to the judgment by a third party (i.e., a garnishee).[9]

Immovable property may similarly be attached in terms of Rule 46, however the process to do so is more onerous. In terms of Rule 46, before a writ is issued for the attachment of immovable property, a further order must be sought from the court declaring the property executable for the judgment debt owing by the judgment debtor.[10] The sheriff is then obliged to attach the property through the physical service of notices of attachment on the owner of the property, the relevant deeds registry and occupier of the property.[11] The sheriff then, upon the request of the execution creditor, sets a date for the sale of the property, which must again by accompanied by a notice of sale physically served on the interested parties.[12] The sale is conducted by public auction and is sold to the highest bidder, with or without a court imposed reserve price.[13] Once the sale is complete, a conveyancer is appointed to attend to the transfer to the purchaser which permits the sheriff to draw up a distribution account for the sale and make payment of the proceeds to the judgment creditor.[14]

---

[4] Rules regulating the conduct of the proceedings of the several provincial and local divisions of the Supreme Court of South Africa as promulgated by Government Notice R48 of 12 January 1965.

[5] Rule 45(1) of the Uniform Rules provides as follows: "45(1) A judgment creditor may, at his or her own risk, sue out of the office of the registrar one or more writs for execution thereof corresponding substantially with Form 18 of the First Schedule."

[6] Rule 45(3) of the Uniform Rules.

[7] Rule 45(7) of the Uniform Rules.

[8] Rule 45(8) of the Uniform Rules.

[9] Rule 45(8) of the Uniform Rules.

[10] Rule 46(1) of the Uniform Rules.

[11] Rule 46(3) of the Uniform Rules.

[12] Rule 46(7) of the Uniform Rules.

[13] Rules 46(10) and 46(12) of the Uniform Rules.

[14] Rules 46(9) and 46(14) of the Uniform Rules.

As can be gleaned from the process described above, relying on traditional methods of execution of judgments is time consuming. For Party A to ultimately obtain satisfaction of its judgment against Party B will take several months if not years, particularly if a writ against movables under Rule 45 remains unsatisfied and a court needs to be approached for an order declaring any immovable property which Party B owns executable. Simply obtaining a sale in execution date from the sheriff can take several months with the actual sale date being some further months into the future.

The process is also costly. For each attendance by the sheriff a fee is charged according to a predetermined court tariff.[15] The sheriff is further entitled to a predetermined commission on the sale value of the assets sold in execution.[16]

To compound the judgment creditor's difficulties, Rule 45A of the Uniform Rules permits a judgment debtor to apply for a stay of execution for such period as a court deems fit.[17] As a general rule the court will grant a stay of execution where an injustice will otherwise result, such as where the underlying basis of the judgment debt is disputed or no longer exists.[18]

In consequence, even post judgment and leaving aside any possible appeals against the judgment handed down, if the judgment debtor is able to raise a *bona fide* dispute to the judgment debt, it is possible to obtain a stay of execution of the order, to the further prejudice of the judgment creditor.

The position is no different should the parties have arbitrated their dispute. Awards issued in arbitration proceedings are not automatically executable. Rather, Sect. 31 of the Arbitration Act[19] provides that an arbitration award may on application by any party to the dispute be made an order of court, which may thereafter be enforced in the same manner as any judgment or order to the same effect.[20]

In our example therefore, Party A would be required to institute a court application, which Party B would be entitled to oppose, seeking that the award in its favour be made an order of court. Only once this has been done, will Party A be able to execute upon the award (now encapsulated in a court order), relying on Uniform Rules 45 and 46.

---

[15] See https://www.sheriffs.org.za/sheriff-fees-tariffs/ Accessed 23 May 2023 for the sheriff's tariffs.

[16] Ibid.

[17] Rule 45A of the Uniform Rules provides as follows: "The court may, on application, suspend the operation and execution of any order for such period as it may deem fit: Provided that in the case of appeal, such suspension is in compliance with section 18 of the Act."

[18] Erasmus: Superior Court Practice RS 16, 2021, at p. D1–604.

[19] Arbitration Act 42 of 1965.

[20] Ibid at section 31, which reads as follows: "Award may be made an order of court (1) An award may, on the application to a court of competent jurisdiction by any party to the reference after due notice to the other party or parties, be made an order of court. (2) The court to which application is so made, may, before making the award an order of court, correct in the award any clerical mistake or any patent error arising from any accidental slip or omission. (3) An award which has been made an order of court may be enforced in the same manner as any judgment or order to the same effect."

Assuming further that Party B owns assets beyond the jurisdiction of South African courts, Party A's position is exacerbated. To attach those assets, Party A would be required to seek the recognition of its high court judgment or arbitral award in that jurisdiction. If the judgment or award is recognised, the traditional means of execution in that jurisdiction would have to follow. Conversely, if the judgment or award is not recognised for whatever reason, there is no further recourse available to Party A in that jurisdiction.

In summary therefore, traditional means of execution as provided for in the Uniform Rules do not lend themselves to swift execution of judgments, often rendering them unattractive to litigants.

If that is so, could on-blockchain execution offer an alternative?

5.3 Smart Contracts

The answer, it is proposed, lies in the use of self-executing 'smart contracts' which operate on a blockchain protocol. Before exploring the proposal in practical terms, it is useful to consider how smart contracts function and what their capabilities are.

The concept of smart contracts originated in 1994, by the publication of Nick Szabo's 'Smart Contracts' paper.[21] Szabo regarded smart contracts as "*a computerised transaction protocol that executes the terms of a contract*".[22]

Whilst there is no unanimity as to the legal definition of smart contracts,[23] the Smart Contracts Alliance has proposed the following useful definition: "*Computer code that, upon the occurrence of a specified condition or conditions, is capable of running automatically according to pre-specified functions. The code can be stored and processed on a distributed ledger and would write any resulting change into the distributed ledger*".[24]

McKinney et al. identify three characteristics that smart contracts share.[25] First, the contract must be 'smart' in the sense that it is represented and executed in computer code (i.e. automated software).[26] Second, a smart contract must automatically give effect to a legally enforceable exchange of promises once the predefined conditions have been met.[27] Third, it must be self-executing, in the sense that once the conditions for the operation of the contract are met, a predefined—and indeed unalterable—execution protocol will be given effect to, and cannot be delayed, halted or reversed.[28]

---

[21] Szabo 1994.
[22] Ibid.
[23] Madir 2018, p 2.
[24] Smart Contracts Alliance 2018.
[25] McKinney et al. 2018.
[26] Ibid at 321.
[27] Ibid at 322.
[28] Ibid at 323.

A smart contract is therefore not a contract in the traditional sense but is rather a computer program that executes a pre-defined transaction upon the happening of certain conditions. An example of this might be a purchase of shares transaction, in which one party sells shares at a defined price, which another party wishes to purchase. A smart contract may be capable of matching the seller's offer with the purchaser's acceptance, and the moment the match takes place the transaction is self-executed by means of the shares being transferred to the purchaser's brokerage account and the seller's brokerage account being credited with the proceeds.

An important feature of smart contracts for purposes of the present argument is that they are built on a blockchain protocol.[29] The advantage of this is that the transactions recorded on the blockchain for purposes of executing smart contracts, cannot be altered.[30] Further, once implemented, a smart contract transaction cannot be halted or reversed (i.e., it is immutable).[31]

Smart contracts can be used for allocating digital currency between two parties upon the requirements of the contract being fulfilled.[32]

The smart contract (i.e., the execution of its code) can be triggered with reference to specified and trusted third party 'off chain' information, making use of blockchain 'oracles'. Oracles are entities that connect blockchains to systems external to the blockchain, enabling smart contracts to execute based upon inputs from the real world.[33]

Koulu explains the concept of smart contract functioning by way of a practical example of a weather bet.[34] In her example two persons place a bet about the weather at a predetermined destination on a specific day. They agree that the winner of the bet will be decided with reference to official weather reports provided by a third party (i.e., the oracle in this example). They create a smart contract and deploy it on, for example, the Ethereum blockchain. Both parties transfer money from their respective accounts to the smart contract and this money is then allocated by the contract to the winner. Based on the third-party input (i.e., verification of the outcome), on the specific day, the smart contract executes and pays the pre-agreed sum without either

---

[29] Koulu 2016, p 53.

[30] Schrepel 2021, p 30. Schrepel explains that: "Blockchain is immutable because it is both decentralized and distributed. The decentralization of blockchain determines who controls it, whilst its distribution designates the location…Blockchain is decentralized since no single user controls the information or data on the blockchain…Put simply, no single user retains the power to alter the information contained in other users' copies. This in alterability applies to blockchain developers, courts and other forms of public intervention."

[31] Schrepel 2021, p 30. Schrepel writes that: "With smart contracts, blockchain users ensure that a set of transactions will be automatically triggered if certain pre-defined conditions are met. Once these potential transactions are put on the blockchain, they cannot be modified or stopped—that is, unless a back door has been implemented to allow designated users to intervene, and as long as the blockchain on which they are implemented is maintained."

[32] Koulu 2016, p 53.

[33] Chainlink 2021.

[34] Koulu 2016, pp 57–64. The example is usefully explained with reference to the source code used to create the smart contract.

party having to initiate the transaction, and without any opportunity for the losing party to renege on the agreement.

5.4 Application in the South African Legal Context

Applying the concept of smart contracts to the question of expedited execution of court orders in South Africa, two primary use cases come to the fore.

5.4.1 Execution

5.4.1.1 Implementation of the Smart Contract

Smart contracts may be harnessed by litigating parties to obviate the need for execution in terms of Uniform Rules 45 and 46, by binding the parties to judgment enforcement through smart contract execution, prior to the litigation having been finalised.

Going back to our example, Party A and B could, at inception of their dispute, agree that the judgment debt will be paid in cryptocurrency through the execution of a smart contract. When exactly this will be agreed depends on the circumstances, save that it will have to be prior to judgment being issued. Likely, if such an agreement can be reached, it will be done at inception of the suit, when summons is served, or the arbitration agreement is signed. (If left too late, a party who felt that the matter was not progressing in its favour may be reticent to agree to smart contract execution).

Each party can open a wallet into which the necessary amount of cryptocurrency for purposes of the suit is transferred (alternatively the amount can be transferred to the smart contract). The parties would then be required to code or have coded a smart contract which, upon notification by the prespecified oracle of the outcome of the suit, would make payment to Party A of an amount in cryptocurrency equal to the judgment debt if judgment is obtained in favour of Party A, or if the suit is dismissed, makes payment to Party B of the security for costs.[35]

Upon notification by the oracle of the final result the smart contract will execute and give effect to whichever of the two scenarios has come to pass. The smart contract will have to make provision for the delivery of notices of appeal or the determination

---

[35] Low 2021, p 6. proposes in the same context that "Most seamlessly, it makes sense for courts to provide template code for the commonly-used blockchain systems." and that "The code itself can be prepared by government vetted professional organisations which establish effective bilingualism in legal drafting and code and can be audited from time to time for completeness, effectiveness and accuracy". Whilst this would no doubt be useful to the parties, aid adoption of on blockchain judgment execution and promote uniformity in smart contract functioning, it is difficult to envisage that the South African courts or government will have the capacity to assist parties in this manner for the foreseeable future.

of any appeals, prior to execution, should the possibility of an appeal be available to the parties.[36]

To be clear, agreeing on this course of action has the result that the usual court sanctioned execution steps under Uniforms Rules 45 and 46 are bypassed completely. What is proposed is ultimately no different to a court order or arbitration award being issued and the losing party complying with the outcome voluntarily.

The main distinction is that, in using a smart contract to give effect to the judgment, such voluntary compliance is built into the process up front, ahead of any order being issued. There is no in principle difficulty with the parties agreeing to this, and for obvious reasons—the attraction of the arrangement is the very fact that whatever the outcome might be, it will be given effect to without either party being burdened with the arduous task of execution under Rules 45 and 46.

Crucially, the parties must be aware that a court or arbitral tribunal will not oversee the process of smart contract agreement or be involved in any manner in its formation, for three reasons. First, a court will not concern itself with execution prior to the outcome of a dispute, save if it is specifically approached to do so (for instance for the issuing of a *Mareva* injunction).[37] Second, neither court functionaries nor judges/arbitrators are likely to have the technical expertise necessary to exercise control over the forming or execution of a smart contract. Third, the parties' arrangement regarding execution will form a separate agreement, which will not be the subject matter of the dispute then pending before the judge/arbitrator and will therefore be ancillary to the main dispute from the presiding officer's perspective.

5.4.2 Practical Impediments to Implementation of Blockchain Execution of Court Orders

While the proposal of smart contract execution outside of the ambit of the Uniform Rules may sound simple in theory, several practical difficulties present themselves.

The first is that the process is by its nature technical and novel. The litigants may need to adjust to the concept of permitting code to execute the result of litigation and be comfortable with trusting the blockchain process to give effect appropriately to the result. They may be put off by the concept of having to code the smart contract required from scratch (whether this is done by the litigants themselves or an external party). Interestingly, it appears that as at February 2021 there are already developments under way to standardise the technology that may in due course permit litigants to use a standardised program with an accessible user interface to code their smart

---

[36] Rule 49(1) of the Uniform Rules in the ordinary course permits fifteen days for the delivery of an application for leave to appeal. If this option is available to the parties (which could be excluded in arbitration) execution will have to be delayed for this period.

[37] A Mareva injunction entails an interim interdict in terms of which a respondent is prevented from disposing of its assets in order to defeat a judgment the applicant believes it will obtain in due course. The order is so structured as to prevent a respondent from placing its assets beyond the reach of the applicant. See Harms 2008, para 426.

contract.[38] If this technology comes to fruition, this impediment to adoption may be substantially reduced.

The second is that a South African court or arbitral tribunal will not issue an order payable in cryptocurrency. Any judgment or award entered will ordinarily be in South African Rands.[39] The smart contract will therefore have to convert the judgment debt expressed in South African Rands, to the cryptocurrency in which payment will be made, on the day of payment. This difficulty is not insurmountable but needs to be taken account of when coding the smart contract.

The third is that any judgment handed down by a court or arbitral tribunal, will be submitted to the parties in written form, either by email or through the South African High Court's case management system.[40] Whatever the order might be, will therefore have to be made available to a human actor who will act as oracle, by submitting the result of the order onto the blockchain for the smart contract to execute. The oracle will have to be an independent person or entity chosen by the parties beforehand. (In due course one might find companies acting as professional oracles in submitting outcomes of judgments/awards onto blockchains to give effect to smart contract execution—akin to escrow agents).

On this point, Ortolani takes the view that: *"The mechanism of oracles can be readily applied to arbitration; a smart contract can defer to the decision of a third party adjudicator, such as an arbitral tribunal, and determine the final recipient of certain disputed assets on the basis of a ruling made by that oracle. In other words, the external information retrieved by the smart contract could be an arbitral award, and software script could be used to enforce the outcome of the procedure."*[41]

Whilst this is no doubt possible in theory, delivering on the promise of seamless interaction between a software script and an arbitral award to execute a smart contract is unlikely to be standard in South Africa for the foreseeable future, requiring the intermediation of a human actor serving as oracle.

---

[38] A United States patent application was lodged on 1 February 2021 under patent number US 10,909,644 B2, seeking to register a patent for 'BLOCKCHAIN-BASED JUDGMENT EXECUTION'. The abstract describes the patent as: "Disclosed herein are methods, systems, and apparatus, including computer programs encoded on computer storage media. One method includes: receiving a request associated with an account of a blockchain-based application for collecting a monetary award issued in an order of a court; determining a creditor, a debtor, and an amount of the monetary award; determining that the account is associated with the creditor based on data recorded on the blockchain; identifying, based on the data, a payment account of the creditor and one or more payment accounts of the debtor with an aggregated balance greater than or equal to the amount of the monetary award; transferring the amount of the monetary award from the one or more payment accounts of the debtor to the payment account of the creditor; and recording a verified time stamp representing a time the amount of the monetary award is transferred." Available at https://www.patentguru.com/US10909644B2 Accessed 23 May 2023.

[39] South African courts have the power to grant a judgment in a foreign currency, however, they will not do so when adjudicating a domestic dispute.

[40] The Gauteng Division and Gauteng Local Division of the High Court uses the CaseLines court management system. See https://caselines.com/solutions/courts Accessed 23 May 2023.

[41] Ortolani 2019, p 439.

The fourth is that the value of cryptocurrencies are notoriously volatile.[42] Where at the inception of a suit the parties may agree to transfer cryptocurrency of an agreed value to be used for smart contract execution, the fluctuations in value of the currency may result in the amount in the wallet being too little to satisfy the order at the time of execution. Conversely, where the value of the currency increases a party might find that it has provided cover for more than any possible judgment amount, yet it may not withdraw any funds from the wallet or smart contract. This can possibly be managed by adopting a time-based review mechanism (say on a three-monthly basis) in which a party will be required to 'top up' the amount available in the wallet or will be entitled to withdraw any excess funds over the amount necessary to sufficiently cover the judgment debt.

The fifth, which emerges from the process described above and the notional difficulties which might arise, is that the parties will have to conclude a written agreement regulating the disparate aspects involved in smart contract execution. The agreement will have to regulate issues such as the creation of the smart contract code, the blockchain platform which will be used, the manner of conversion of South African Rands to cryptocurrency, the way the outcome of the suit will be submitted onto the blockchain, the oracle to be used and ensuring that the smart contract retains a sufficient value of cryptocurrency to give effect to any order entered against the parties. The concept of delayed execution for appeals and the possible stay of execution in prescribed circumstances will also need to be catered for. Such an agreement will be detailed and technical in nature and will likely form the subject matter of much negotiation between the parties. The agreement itself may also become the subject of dispute between the parties. These elements of complexity render the adoption of such an agreement relatively unattractive. It may be however that in due course a standard form of agreement can be adopted should blockchain execution become more widely practised.

5.4.3 Benefits to the Parties

A seemingly simple yet fundamental question that must be asked is why the parties would agree to blockchain execution at all? The advantages from the claimant's perspective are clear. Any order obtained will be executed upon immediately and the cumbersome traditional execution process negated. The process is automated on-blockchain and therefore free from human interference. The claimant would be guaranteed payment and a recalcitrant debtor will not have the ability to evade satisfaction of the order. Once the smart contract executes, the payment is final and cannot be reversed. The result will therefore be immutable. The process is swift and cost effective. All these factors redound to the claimant's benefit.

---

[42] See for instance https://www.bankrate.com/investing/bitcoin-and-crypto-crash-what-investors-should-do/.

From the defendant's perspective, the advantage is less clear. Even a defendant acting in good faith might be hesitant to commit a substantial sum of cryptocurrency to a smart contract for an extended period whilst litigious proceedings are ongoing, simply to suit the convenience of a claimant. Such a defendant might take the view that when judgment is handed down, it will comply with the judgment in the ordinary course.

There are however two scenarios in which a defendant may well be inclined to agree to be bound upfront to on-blockchain execution. The first is where the defendant wishes to institute a counterclaim against the claimant. The second is where the defendant wishes to obtain security for costs (explored more fully below).

Counterclaims in South African law are provided for in Uniform Rule 24, which permits a defendant to deliver a claim in reconvention against the claimant, at the time of pleading to the claimant's suit.[43] A counterclaim may be instituted conditionally (i.e. on the basis that it will only be activated in the event that the claimant's suit fails).[44] Counterclaims may of course also be instituted in arbitration and are common in both litigation and arbitration, particularly in commercial disputes where both parties have suffered a loss arising from a specific set of circumstances, for instance as a result of an infrastructure project not being completed or an external event leading to a declaration of *force majeure* by one of the parties. In such an event, both parties may wish to lay the blame at each other's doors and deliver both claims in convention and reconvention.

In such a scenario, where Party B might wish to institute a counterclaim against Party A, the prospect of both parties committing to on-blockchain execution would make sense to both litigants—both parties will be claimants in the dispute and benefit from receiving immediate redress by way of the pre-agreed smart contract execution.

This is particularly so where time is of the essence to both parties. For instance, the parties may choose to arbitrate their dispute using the Arbitration Foundation of South Africa's Expedited Rules, which are expressly aimed at ensuring that the dispute is speedily resolved.[45] If an award is issued expeditiously in an expedited

---

[43] Rule 24(1) of the Uniform Rules provides that: "24(1) A defendant who counterclaims shall, together with his plea, deliver a claim in reconvention setting out the material facts thereof in accordance with rules 18 and 20 unless the plaintiff agrees, or if he refuses, the court allows it to be delivered at a later stage. The claim in reconvention shall be set out either in a separate document or in a portion of the document containing the plea but headed 'Claim in Reconvention'. It shall be unnecessary to repeat therein the names or descriptions of the parties to the proceedings in convention."

[44] Rule 24(4) of the Uniform Rules provides that: "24(4) A defendant may counterclaim conditionally upon the claim or defence in convention failing."

[45] AFSA 'Rules for Expedited Arbitration' (October 2021). Available at https://arbitration.co.za/domestic-arbitration/expedited-rules/ Accessed 23 May 2023. Rule 6(5) makes provision for expedited proceedings: "6.5 The ARBITRATOR has the widest discretion and powers allowed by law to ensure that the just, expeditious, economical and final determination of all the disputes raised in the proceedings including the matter of costs and, if needs be, he/she shall have all the powers accorded to an ARBITRATOR acting under the AFSA Rules for Administered Arbitrations. All powers and functions exercised by the ARBITRATOR shall be in accordance with the provisions of the Arbitration Act of 1965."

arbitration, there is little sense in the parties having to struggle thereafter to execute in the ordinary course—an expedited arbitration inextricably married to blockchain execution would ensure that the entire process is concluded within weeks, to the benefit of the claimant in both the main claim and counterclaim.

5.4.4 Security for Costs

A second basis upon which a defendant is likely to agree to on-blockchain execution, is where the defendant stands to gain from such execution in respect of security for costs. In the scenario detailed above, Party A was required to render security for the costs of the litigation to Party B. Security for costs is furnished in terms of Uniform Rule 47, which permits a litigant to seek security for the costs which it may incur in litigation from the other party to the suit.[46] Rule 47(5) provides that security for costs must "*...be given in the form, amount and manner directed by the [High Court] registrar.*"[47]

Whilst the Rule does not prescribe which form security for costs must take, it is established practice that it takes the form of a bank guarantee.[48] This entails the litigant required to put up security, to deposit in a bank account the amount at which security was set. The bank then issues a physical guarantee for the amount held by it. The guarantee undertakes to make payment to the litigant requiring security that it will pay to such litigant the amount held in the account, upon the happening of certain events, usually the presentation of a finalised bill of costs setting out the amount payable. When the guarantee becomes payable, it must be physically presented at a bank branch for payment.

As is clear, the process of obtaining such a guarantee is (like traditional execution) a laborious and expensive undertaking. It requires payment to the bank concerned of a fee or commission to issue the guarantee and maintain the account in which the funds are held. Also, whilst the funds are held by the bank subject to the guarantee, the interest earned on the amount deposited is usually of a nominal nature, leading to a depreciation of the amount held over time.

This being so, it may be attractive for a defendant (Party B in our scenario), to agree to on-blockchain execution and the putting forward of a possible judgment amount as part of the smart contract, provided that the claimant (Party A in our scenario) is willing to do the same in respect of any security for costs agreed to between the parties. Party B will in this event enjoy the certainty of knowing that

---

[46] Rule 47(1) of the Uniform Rules provides that: "47(1) A party entitled and desiring to demand security for costs from another shall, as soon as practicable after the commencement of proceedings, deliver a notice setting forth the grounds upon which such security is claimed, and the amount demanded."

[47] Rule 47(5) of the Uniform Rules provides that: "47(5) Any security for costs shall, unless the court otherwise directs, or the parties other-wise agree, be given in the form, amount and manner directed by the registrar."

[48] Erasmus: Superior Court Practice RS 16, 2021, D1-648.

no bank guarantee is required nor need it be presented for payment if the claimant's suit is successfully dismissed—rather, the dismissal will be actioned on-blockchain and the costs due to the defendant paid to it by means of pre-agreed smart contract execution. Party B may particularly be inclined to agree to this where it believes the claimant's suit has little prospect of success.

5.5 Conclusion

It is clear from what is set out in this chapter that conventional means of execution in South Africa holds little appeal to a litigant with an order in its favour. The process is, as explained, time consuming and expensive.

An alternative to the *status quo* could be on-blockchain execution using smart contracts. The benefits to the parties are evident—speed of execution and certainty of payment is a boon to any litigant.

This is particularly so where both parties stand to gain, such as where both a claim and counterclaim is concerned, or either party stands to profit from security for costs put up in litigation.

While the proposal is novel, certainly in a South African context, as cryptocurrencies and blockchain technology become more widely used and gain legitimacy in commerce, so could the use of smart contract execution, to the benefit of the litigious process as a whole.

References

Chainlink (2021) What is a blockchain oracle? https://chain.link/education/blockchain-oracles Accessed 23 May 2023

Harms DR (2008) The Law of South Africa, Interdict, Volume 11, 2nd edn. Lexis Nexis South Africa, Durban

Harms DR (2016) The Law of South Africa, Civil Procedure: Superior Courts, Volume 4, 3rd edn. Replacement. Lexis Nexis South Africa, Durban

http://www.fon.hum.uva.nl/rob/Courses/InformationInSpeech/CDROM/Literature/LOTwintersch ool2006/szabo.best.vwh.net/smart.contracts.html Accessed 23 May 2023

Koulu R (2016) Blockchains and Online Dispute Resolution: Smart Contracts as an Alternative to Enforcement. Scripted vol. 13, issue 1. SCRIPTed, Edinburgh

Low ZE (2021) Execution of judgments on the blockchain: A practical legal commentary. Harvard Journal of Law & Technology, vol. 34, Spring 2021. Harvard Law School, Cambridge

Madir J (2018) Smart Contracts: (How) Do They Fit Under Existing Legal Frameworks? Social Science Research Network, Rochester

McKinney SA, Landy R, Wilka R (2018) Smart Contracts, Blockchain and the Next Frontier of Transactional Law. Washington Journal of Law, Technology & Arts, vol. 13, issue 3. University of Washington School of Law, Seattle

Ortolani P (2019) The impact of blockchain technologies and smart contracts on dispute resolution: arbitration and court litigation at the crossroads. Uniform Law Review vol. 24, issue 2. Oxford University Press, Oxford

Schrepel T (2021) Blockchain + Antitrust - The Decentralization Formula. Edward Elgar Publishing, Cheltenham

Smart Contracts Alliance (2018) Smart Contracts: Is the Law Ready? https://digitalchamber.org/smart-contracts-paper-press/ Accessed 23 May 2023

Szabo N (1994) Smart Contracts. Available at http://www.fon.hum.uva.nl/rob/Courses/InformationInSpeech/CDROM/Literature/LOTwinterschool2006/szabo.best.vwh.net/smart.contracts.html Accessed 23 May 2023

Chapter 6
The DLT Pilot Regime and DeFi

Guilherme Maia and João Vieira dos Santos

Contents

6.1	Background of the DLT Pilot Regime	90
6.2	What is the DLT Pilot Regime?	91
6.3	What Is DeFi?	96
	6.3.1 DeFI Architecture	97
	6.3.2 DeFi Categories	99
6.4	How Does the DLT Pilot Regime Match with DeFi?	102
References		105

Abstract This chapter discusses the new DLT (Distributed Ledger Technology) Pilot Regime issued by the EU, considering its use and applicability regarding DeFi (Decentralized Finance). As set out in the ESMA (European Securities and Markets Authority) Report on crypto-assets, the authorities are having difficulties interpreting the existing legal requirements, discussing their application to crypto-assets and DLT structures. The DLT Pilot Regime is intended to operate in a way akin to a regulatory sandbox, permitting and framing the operation of market infrastructures that process trades in crypto-assets qualified as financial instruments and that otherwise would not be allowed to operate or would be subject to excessive and burdensome requirements. The regime aims to create legal certainty in view of the rules currently defined for secondary markets in financial instruments, particularly if these rules are fully suited to DLT and crypto-assets while also ensuring the protection of investors, market integrity and financial stability.

Keywords DLT · DLT pilot regime · Blockchain · Capital markets union · Crypto

---

G. Maia (✉)
BCAS, Floor 1, Mompalao Buildings, Tower Street, Msida, MSD 1825, Msida, Malta
e-mail: guilhermeinvestimento@gmail.com

J. Vieira dos Santos
Lusófona University and Portuguese Securities Market Commission (CMVM), R. Laura Alves 4, 1050-124 Lisbon, Portugal
e-mail: p6583@ulusofona.pt

© T.M.C. ASSER PRESS and the author(s) 2024
F. Pereira Coutinho et al. (eds.), *Blockchain and the Law*, Information Technology and Law Series 37, https://doi.org/10.1007/978-94-6265-579-9_6

6.1 Background of the DLT Pilot Regime

The European Commission on 24 September 2020 adopted a new package on "Digital Finance", which includes digital finance and retail payments strategies, as well as legislative proposals on crypto-assets and digital resilience.[1] The legislative proposals on crypto-assets were based on a public consultation conducted between December 2019 and March 2020[2] and on the reports from ESMA (European Securities and Markets Authority)[3] and EBA (European Banking Authority)[4] about crypto-assets. In those reports, two main needs were identified.

The first need concerns introducing rules in an area that is not yet regulated, concerning crypto-assets that do not qualify as other instruments of the European Union's financial legislation. The second need is related to adapting some rules applicable to crypto-assets that qualify as financial instruments to respect the principle of technological neutrality and not create obstacles to innovation.

For the first case, the European Commission has presented a proposal for a European Regulation on Markets in Crypto-assets, better known as MiCA.[5] For the second case, a proposal has been made by the European Commission for a pilot regime for market infrastructures based on distributed ledger technology.

This last proposal intends to implement an approach similar to that of a regulatory sandbox with a more specific focus on only trading in crypto-assets qualified as financial instruments on market infrastructures.[6] This aims to create legal certainty before the rules currently defined for secondary markets in financial instruments, particularly if these rules are fully suited to distributed ledger technology and crypto-assets while ensuring the protection of investors and market integrity and financial stability.

As set out in the ESMA Report on crypto-assets, the authorities encounter difficulties interpreting the existing requirements when they qualify as financial instruments. Some rules are unsuitable for distributed ledger technologies (hereinafter also referred to as DLT) and are mainly related to the secondary market. This understanding has some correspondence in the market, where several public offerings of crypto-assets qualified as financial instruments have already been made; however, experiences in the secondary market are much scarcer.

Addressing these setbacks in the secondary market for crypto-assets qualified as financial instruments could boost the primary market by ensuring more liquidity for

---

[1] Available at: https://ec.europa.eu/info/publications/200924-digital-finance-proposals_en. Accessed 3 May 2023.

[2] Available at:
https://ec.europa.eu/info/sites/default/files/business_economy_euro/banking_and_finance/documents/2019-crypto-assets-consultation-document_en.pdf. Accessed 3 May 2023.

[3] Available at: https://www.esma.europa.eu/document/advice-initial-coin-offerings-and-crypto-assets. Accessed 3 May 2023.

[4] Available at: https://www.eba.europa.eu/eba-reports-on-crypto-assets. Accessed 3 May 2023.

[5] See in this respect, among others, Maia and Santos 2022; Zetzsche et al. 2020a.

[6] Convergently, Zetzsche and Woxholth 2021, p 4.

these instruments issued and traded through distributed registration technologies. The proposal itself explicitly addresses this issue by stating in one of its recitals the following: "Without a secondary market able to provide liquidity and to enable investors to buy and sell such assets, the primary market for crypto-assets that qualify as financial instruments will never expand in a sustainable way".

The negotiations on this proposal have ended, and on 2 June 2022, the Regulation (EU) 2022/858 of the European Parliament and the Council (hereinafter referred to as DLT Pilot Regime) on a pilot regime for market infrastructures based on distributed registry technology and amending Regulations (EU) No 600/2014 and (EU) No 909/2014 and Directive 2014/65/EU was published.

6.2 What is the DLT Pilot Regime?

The DLT Pilot Regime allows market infrastructures using distributed ledger technologies to be temporarily exempted[7] from certain specific requirements under European Union financial legislation that might otherwise prevent them from developing solutions for the trading and settling transactions in crypto-assets that qualify as financial instruments.

Additionally, it is an optional regime, not preventing financial market structures, such as multilateral trading facilities (MTFs), organized trading facilities (OTFs), central securities depositories (CSDs) and central counterparties (CCPs), from carrying out their trading and post-trading activities for crypto-assets that qualify as financial instruments under EU financial legislation. The DLT Pilot Regime also allows ESMA and competent authorities to gain experience on the specific opportunities and risks of distributed ledger technologies in the trading and post-trading services.

The scope of the DLT Pilot Regime is limited to the conditions applicable to the operation of market infrastructures based on DLT, the authorizations for their use and the supervision and coordination of competent authorities and ESMA. In terms of a subjective matter, the Regulation applies to investment firms, market operators and CSDs.

Only such entities may operate DLT market infrastructures, with one important exception: new entrants. It was determined that access to the DLT Pilot Regime should not be limited to incumbents, so an entity that is not authorized under Regulation (EU) No 909/2014 or Directive 2014/65/EU may apply for an authorization under that Regulation or Directive, respectively, and at the same time for a specific authorization under the DLT Pilot Regime—Articles 8(2), 9(2) and 10(2) of the DLT Pilot Regime.

In such cases, the national competent authority should not assess whether that entity meets the requirements of Regulation (EU) No 909/2014 or Directive 2014/

---

[7] Specific authorizations and exemptions should be of a temporary nature, for a maximum period of six years from the date the specific authorization is granted and should only be valid for the duration of the DLT Pilot Regime—Articles 8(11) and 9 (11) of the DLT Pilot Regime.

65/EU for which an exemption under the DLT Pilot Regime has been requested. This means those entities are legal persons with limited authorizations as they can only operate DLT market infrastructures in accordance with the DLT Pilot Regime. Their authorization should be withdrawn upon the expiry of their specific authorization unless the entities submit a complete application for authorization to be a CSD under Regulation (EU) No 909/2014 or an investment firm or a market operator under Directive 2014/65/EU.

The DLT Pilot Regime introduces this new status regarding the DLT market infrastructures—Article 2, point 5, of the DLT Pilot Regime. It includes DLT MTFs, DLT SSs (settlement systems) and DLT TSSs (trading and settlement systems). Also important for this new status was the introduction of another concept, that of DLT financial instruments, which correspond to financial instruments that are issued, recorded, transferred, and stored through a distributed ledger technology—Article 2, point 11, of the DLT Pilot Regime.[8]

Therefore, DLT MTFs are MTFs that only admit to trading DLT financial instruments (Article 2, point 6 of the DLT Pilot Regime). On the other hand, DLT SSs are settlement systems that settle transactions on DLT financial instruments, irrespective of the fact that these settlement systems have been designated and notified under Directive 98/26/EC (so as not to be limited to the definition of "system" of this Directive), and that allow either the initial recording of financial instruments DLT or the provision of safekeeping services in relation to DLT financial instruments—Article 2, point 7, of the DLT Pilot Regime. Finally, DLT TSSs are DLT MTFs or DLT SSs that combine the services provided by a DLT MTF and a DLT SS—Article 2, point 10, of the DLT Pilot Regime.

As this is an experimental regime, the following limitations are foreseen as to the type and volume of admitted financial instruments—Article 3(1) of the DLT Pilot Regime:

- Shares whose issuer has a market capitalization or provisional market capitalization of less than € 500 million;
- Bonds, other forms of securitized debt or money market instruments (except derivatives or complex products) with an issue size below € 1 billion;
- Units in collective investment undertakings whose market value of assets under management is less than € 500 million.

The aggregate market value of all the DLT financial instruments admitted to trading or recorded in a DLT market infrastructure may not exceed € 6 billion, and if the admission to trading or the initial recording of a new DLT financial instrument results in an aggregate market value reaching that value the DLT market infrastructure does not admit that DLT financial instrument to trading or registration—Article

---

[8] DLT financial instruments can also be said to be crypto-assets that qualify as financial instruments, as the MiCA defines "crypto-asset" as a digital representation of value or rights that can be transferred and stored electronically using distributed ledger technology or similar technology. The DLT Pilot Regime also took the opportunity to clarify that financial instruments issued using distributed ledger technology fall within the definition of "financial instrument" in Directive 2014/65/EU by including an amendment to Article 4.

6 The DLT Pilot Regime and DeFi

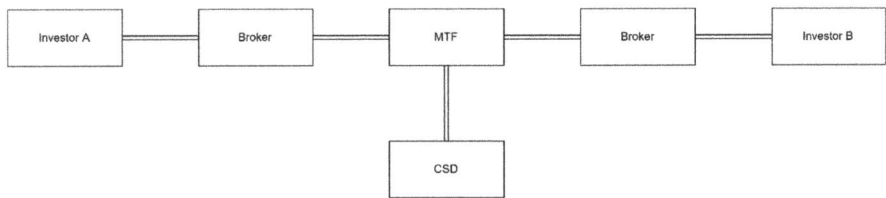

Fig. 6.1 Traditional Market Operation. *Source* The authors

3(2) of the DLT Pilot Regime. However, all the above-mentioned thresholds are subsidiary, considering that the national competent authorities may establish lower thresholds—Article 3, paragraph 6, of the DLT Pilot Regime.

Suppose the aggregate market value of all DLT financial instruments admitted to trading on a DLT market infrastructure or registered on a DLT market infrastructure has reached € 9 billion. In such case, the operator of the DLT market infrastructure activates the transition strategy and notifies the national competent authority—Article 3(3) of the DLT Pilot Regime.

Regarding exemptions granted to operators of DLT market infrastructures, DLT MTFs can be exempted from the intermediation obligation under Directive 2014/65/EU. Traditional MTFs are only allowed to admit as members or participants investment firms, credit institutions and other persons with a sufficient level of trading capacity and expertise and who maintain adequate resources and organizational arrangements. Therefore, a traditional secondary market transaction can be presented as can be seen in Fig. 6.1 below:

In the above diagram, we have included the brokers that perform the intermediation referred to above, being the entities responsible for the reception and transmission of the investors' orders to the MTF. It is necessary for an entity that performs its management and a CSD that manages the settlement system.

However, crypto-assets trading platforms provide direct access to retail investors, not least because of the greater security that distributed ledger technologies provide for the circulation of value in a digital medium. For this reason, the DLT Pilot Regime allows the competent authority to grant, upon request by an operator of a DLT MTF, a temporary exemption from that intermediation obligation to provide direct access to retail investors and allow them to trade on their own accounts.[9] That said, we present a transaction on a DLT MTF (see Fig. 6.2) where such an exemption has been granted, as follows:

Brokers are not required due to the direct access of investors to the DLT MTF. As a result, we only find two service providers (the operator of the DLT MTF and the CSD) in a market transaction. Certain conditions are defined for retail investors to have direct access to a DLT MT, which need to be assessed by the national competent authority. In addition, those authorities may require additional investor protection measures from the operators of DLT MTF—Article 4(2) of the DLT Pilot Regime.

---

[9] The system of holding indirectly through a financial intermediary presents obvious problems related to the multiplicity of intermediaries involved, *vd.* Correia 2017, p 72.

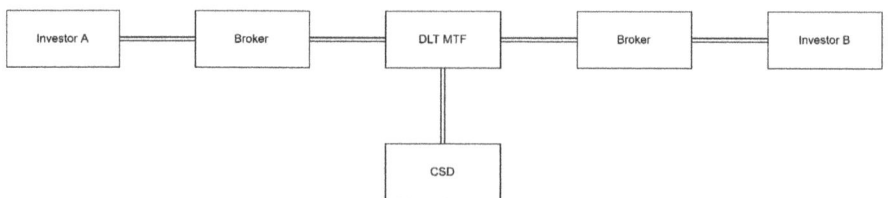

Fig. 6.2 DLT MTF operation. *Source* The authors

Pursuant to Article 5 of the DLT Pilot Regime, a CSD aiming to operate a DLT SS may apply to national competent authorities, subject to appropriate reasoning, for the following exemptions from Regulation (EU) No 909/2014 *(inter alia)*:

- Article 2(1), subparagraphs (4), (9) and (28)—definitions of dematerialized form, transfer order and securities account;
- Articles 3, 37 and 38—recording in book-entry form, the integrity of the issue and protection of securities;
- Articles 6 and 7—measures to prevent and address settlement fails;
- Article 19–extension and outsourcing of activities and services;
- Articles 33, 34 and 35—requirements for participation, transparency, and communication procedures;
- Articles 39 and 40—settlement finality and cash settlement;
- Articles 50, 51 and 53—access between CSDs.

Within the referred rules, we highlight two specific exemptions. We start with the exemption to the obligation of intermediation through a credit institution or an investment firm, which allows retail investors direct access to settlement systems operated by a DLT SS. The second exemption relates to cash settlement entities (in central bank money) to develop innovative solutions under the DLT Pilot Regime, facilitating access to commercial bank money or using «electronic money tokens».[10]

With regard to operators of DLT TSSs, they can apply for the same exemptions available to both operators of DLT MTFs and DLT SSs, provided they comply with the conditions attached to the exemptions and any compensatory measures required by the competent national authorities—Article 6 of the DLT Pilot Regime. That said, we present an operation on a DLT TSS where all the exemptions have been granted, as can be seen in Fig. 6.3 below:

Since natural persons can participate directly in the market and settlement system, brokers are not required. In addition, the DLT Pilot Regime allows the same entity to manage the trading and settlement systems. The combination of exemptions thus allows for only one entity to provide services in the market, coming close to what is intended by the current players from the market on crypto-assets.

---

[10] "Electronic money token" is a category of crypto-assets established in MiCA, defined as a type of crypto-asset the main purpose of which is to be used as a means of exchange and that purports to maintain a stable value by referring to the value of a fiat currency that is legal tender.

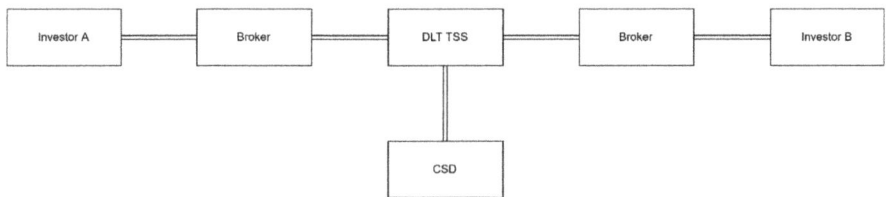

Fig. 6.3 DLT TSS operation. *Source* The authors

The DLT Pilot Regime also provides for a set of additional requirements for operators of DLT market infrastructures, namely: (i) liability in the event of loss of funds, guarantees or a DLT financial instrument; (ii) protection measures in the event of insolvency; (iii) specific reporting duties; and (iv) operational risk management—Article 7 of the DLT Pilot Regime.

The additional requirements are necessary to avoid risks related to distributed ledger technology or how the DLT market infrastructure would operate. An operator of a DLT market infrastructure is, for example, required to inform members, participants, issuers, and clients how it intends to carry out its activities and how the use of distributed ledger technology differs from the way services are typically provided by a traditional MTF or a CSD operating a securities settlement system.

At the time of granting a specific authorization, operators of DLT market infrastructures should also have a credible exit strategy in case the DLT Pilot Regime is discontinued, the specific authorization or some of the exemptions granted are withdrawn, or the thresholds for DLT financial instruments are exceeded. Such strategy should include how to transition or revert from their operations based on distributed ledger technology to traditional market infrastructures—Article 7(7) of the DLT Pilot Regime.

National competent authorities are those designated under Article 67 of Directive 2014/65/EU, under Article 11 of Regulation (EU) No 909/2014 or otherwise by the Member States—Article 2(21) of the DLT Pilot Regime. These authorities provide ESMA with all relevant information before granting a specific authorization—Articles 8(6), 9(6) and 10(7) of the DLT Pilot Regime. This is because ESMA may provide the national competent authority with a non-binding opinion on the exemptions requested or on the appropriateness of the type of distributed ledger technology used, where necessary to promote consistency and proportionality of the exemptions or where necessary to ensure investor protection, market integrity and financial stability—Articles 8(7), 9(7) and 10(8) of the DLT Pilot Regime.

The national competent authority examining an application submitted by an operator of a DLT market infrastructure has the possibility to refuse to grant a specific authorization if there are reasons to believe that the DLT market infrastructure would not be able to comply with applicable provisions established by the Union law or provisions of national law falling outside the scope of Union law, if there are reasons to believe that the DLT market infrastructure would pose a risk to investor protection, market integrity or financial stability, or if the application constitutes an attempt to circumvent existing financial requirements—Articles. 8(10), 9(10) and 10(10) of the DLT Pilot Regime.

After the specific authorization has been granted, competent authorities have the power to withdraw a specific authorization or any exemptions granted to a DLT market infrastructure where an anomaly has been detected in the underlying technology or in the services provided and activities undertaken by the operator of the DLT market infrastructure if that anomaly is not outweighed by the benefits provided by the service and activities in question, where the operator of the DLT market infrastructure has breached any obligations attached to authorizations and exemptions granted by the competent authority, or where the operator of the DLT market infrastructure has registered financial instruments which exceed the thresholds for DLT financial instruments or which do not comply with other conditions applying to such instruments provided for in the DLT Pilot Regime—Articles 8(12), 9(12) and 10(12) of the DLT Pilot Regime.

ESMA must submit a report to the European Commission by 24 March 2026, which includes a cost–benefit analysis in order to determine the extension of the DLT Pilot Regime for a new period of up to three years, the extension of the regime to other types of financial instruments, its modification, its conversion into a permanent regime or its termination—Article 14 of the DLT Pilot Regime. The DLT Pilot Regime entered into force on 22 June 2022, with most of the rules being applicable as of 23 March 2023—Article 19 of the DLT Pilot Regime.

To implement this Regulation is important to mention that the rules on securities recording systems are not harmonized at the EU level,[11] so the DLT Pilot Regime does not directly target such control systems. In this context, the concentration of trading and settlement functions in a single entity, as exemplified in the last diagram above, may have an important link with national rules on recording systems.

6.3 What Is DeFi?

DeFi (Decentralized Finance) is an ecosystem of decentralized applications (dapps)[12] that provide financial services built on peer-to-peer and trustless networks, meaning they do not need a central authority,[13] which began to be relevant in size in 2020.[14] DeFi is usually described as an open, permissionless, and highly interoperable

---

[11] *Vd.* Zetzsche and Woxholth 2021, p 14.

[12] "dapps are programs running on a decentralised network that can execute automatically when certain conditions are met, where transactions are affected in a secure and verifiable way, and legitimate state changes persist on a public blockchain."—Ethereum Foundation, 2020, "What are dapps?" [Online], available at: https://ethereum.org/en/dapps/#what-are-dapps. Accessed 29 April 2021.

[13] Katona 2021, p 76.

[14] "The value of funds that are locked in DeFi related smart contracts recently reached USD 10 billion", Schär 2021, p 2.

6 The DLT Pilot Regime and DeFi   97

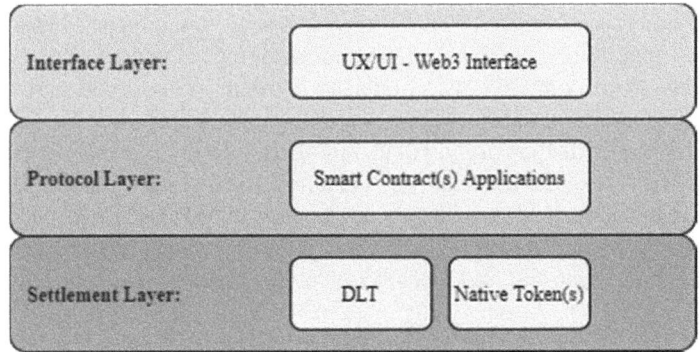

Fig. 6.4 DeFi layered system. *Source* The authors

protocol stack built on public distributed ledger technologies to replicate existing financial services more transparently and openly.[15]

DeFi may comprise a provision of financial services by multiple participants, intermediaries, and end-users across various jurisdictions, with a purely technological infrastructure to facilitate interactions between agents.[16] *"As blockchain technology matures and evolves to process a greater number of transactions, the technology could provide a platform on which to run code-based systems that are increasingly untethered from human control"*.[17]

In this chapter, we give an overview of the DeFi ecosystem by describing the technical infrastructure of the technology and the different applications within the financial industry.

6.3.1 DeFI Architecture

This sub-chapter illustrates a simplified version of DeFi's structure by providing a general and non-technical overview of its layers. DeFi uses a hierarchical multi-layered architecture with different purposes. It differentiates between three main layers, as is shown in Fig. 6.4: the settlement, the protocol, and the interface layers.[18]

I. **The Settlement Layer (Layer One):**

---

[15] Ibid. p 1.
[16] Zetzsche et al. 2020b, p 4.
[17] De Filippi and Wright 2018, p 147.
[18] For a technical conceptual framework based on a OSI Model, please see Schär 2021, p. 4. The proposed framework in this Schär 2021, p. 4 was strongly influenced by the Fabian Schär model, but for reasons of simplicity we have opted for a different approach.

Layer One consists of the DLT[19] and its native asset, containing the basic operating rules of the ecosystem.[20] Ideally, the DLT comprises the hardware layer where a *peer-to-peer* network of computers is required to compute transactions and store them in order in a distributed database. In a blockchain,[21] each agent in the network is a "node", which validates and organizes incoming transactions into blocks broadcasted to the network. These transactions may contain value and information. The value is expressed in the network's native asset, and the information is code that can pass data and trigger actions. The native token is used to transfer value, be used as a means of exchange, run applications, and incentivize "nodes" to maintain neutrality.[22]

II. **The Protocol Layer (Layer Two)**:

Layer Two comprises the compiler and the ability to create Application Programming Interfaces (API). A compiler is a program that converts current high-level to low-level programming languages.[23] In this layer, developers may write code, compile it into bytecode (machine language) and deploy it onto the DLT. Additionally, developers may create an API allowing other developers to interact with the deployed code. An API is a set of definitions and protocols for building and integrating application software that lets a product or service communicate with other products or services without understanding how they are implemented.

Layer Two includes the programming languages that may be compiled, such as Solidity or Python, smart contract[24] standards, and any assets issued on top of the settlement layer. Furthermore, all standards used for specific applications, such as those described in Sect. 6.3.2 and aggregators that may connect several applications and protocols, are included.

III. **The Interface Layer (Layer Three)**:

Layer Three creates user-oriented applications that allow users to interact with an application through a web page. The purpose of interfaces is to enable the program

---

[19] DLT is used in Fig. 6.4 for a technologically agnostic approach. However, for "decentralised finance" to be effective it is required for a DLT to be decentralised by default. For our study, decentralisation is only achieved if (1) the architecture of the system is distributed (2) and if no individual or entity controls the system. The distribution of network governance is only possible if there is a distributed architecture. For a better understanding see https://medium.com/@VitalikButerin/the-meaning-of-decentralization-a0c92b76a274. Accessed 2 May 2023.

[20] Katona 2021, p 81.

[21] As defined by Dr. Keir Finlow-Bates, *"Blockchain comprises a time-stamped hash-linked list distributed over a peer-to-peer network with a consensus algorithm for incentivizing data consistency and utilizing asymmetric key cryptography for identity and access management."* See Finlow-Bates 2020, p 16. Blockchain is the most used sub-category of DLTs for DeFI applications according to https://defipulse.com/. Accessed 2 May 2023.

[22] For a better understanding, see https://medium.com/@micheledaliessi/how-does-ethereum-work-8244b6f55297. Accessed 2 May 2023.

[23] Perumal et al. 2018, pp 100-104.

[24] Regarded herein as self-executing programmes, with no human interaction and carried through computer-running codes, see Raskin 2017, p 306.

to enforce the properties embedded in the code via objects familiar to most users, such as text boxes, buttons, or any interaction on a web page.[25]

6.3.2 DeFi Categories

This sub-chapter aims to describe the current major applications of DeFi. There are categories termed "stablecoins" and "derivatives", which formally refer to the representation of a particular instrument; however, the scope of this sub-chapter is to determine the applications at a protocol level. As a result, those terms are merely used for technical ease of reference.

Currently, DeFi has seven major classes:[26]

IV. **Stablecoins**: Stablecoins are crypto-assets that typically transact on a distributed ledger and rely on cryptographic validation techniques to be transacted, intending to achieve stable value relative to fiat currencies. In principle, stablecoins allow users to protect the nominal value of their holdings.[27] Fundamentally, there are three types of stablecoins: fiat-backed stablecoins, asset-backed stablecoins and algorithmic stablecoins.

V. **Fiat-backed stablecoins** aim to maintain a stable value by being collateralized by the fiat currency that they may represent.[28]

VI. **Asset-backed stablecoins** aim to maintain a stable value via collateralization of several fiat currencies, one or more commodities, other crypto-assets, or a basket of such assets.[29]

VII. **Algorithmic stablecoins** aim to maintain a stable value via inbuilt stabilization functions that provide for the increase or decrease of the supply of such crypto-assets in response to changes in demand.[30]

Decentralized stablecoins (b., c.) aim to solve the trust issue that may arise from fiat-backed stablecoins (a.). Ideally, decentralized stablecoins are created in a decentralized manner via an over-collateralization or algorithmic method, operate fully on decentralized ledgers, are governed by decentralized autonomous organizations, and anyone can publicly audit their reserves. Thus, one can conclude that the core components of a decentralized stablecoin are as follows:[31]

- **Collateral**: This is the store of primary value for a stablecoin. The collateral per se can be exogenous and primarily used in other protocols; endogenous, where

---

[25] https://www.cs.utah.edu/~germain/PPS/Topics/interfaces.html. Accessed 2 May 2023.
[26] Lau et al. 2020, pp 13-16.
[27] https://www.bis.org/cpmi/publ/d187.pdf. Accessed 3 May 2023.
[28] https://tether.to/wp-content/uploads/2016/06/TetherWhitePaper.pdf. Accessed 3 May 2023.
[29] https://makerdao.com/en/whitepaper/#notes. Accessed 3 May 2023.
[30] https://www.ampleforth.org/papers/. Accessed 3 May 2023.
[31] Klages-Mundt et al. 2020, pp 59–79.

the collateral was created to be collateral within the protocol or implicit, where the design lacks an explicit collateral store.
- **Agents**: The collateral providers and the stablecoin users.
- **Governance**: The functions and parameters that govern the protocol.
- **Issuance**: A mechanism to control the issuance of stablecoins.
- **Oracles**: A mechanism to import external data onto the blockchain, such as price feeds.
- **Lending and Borrowing Markets**: These protocols aim to provide lending and borrowing services to anyone who collateralizes their crypto-assets and uses them to obtain loans. Lenders can also earn a yield on their assets by contributing to lending pools and earning interest in such assets.[32] However, these markets are not comparable to peer-to-peer lending since the crypto-assets are not directly lent between individual agents but are borrowed against the smart contract reserves: crypto-funds aggregated and pooled together by lending agents.

Due to the nature of DeFi and blockchain, where borrowers are pseudo-anonymous, there are no mechanisms to identify the agent, access his credit score and ability to repay the loan, and enforce any legal proceedings against a default. To prevent borrowers from defaulting on their debt obligation and creating a credit risk for the loaners, borrowers must collateralize their position to cover the debt value. Due to the historical volatility and illiquidity of the various crypto-assets, an over-collateralization is required. If the underlying asset's market price crashes, it will not affect the lending market funds since sufficient collateral would cover the borrower's debt value. The debt is usually issued against USD or stablecoins representing USD.

The borrower must ensure that the collateral value is always above a pre-determined liquidation threshold set by the protocol, or the "liquidators" can purchase the locked collateral at a discount and close the borrower's debt position.[33]

IX. **Exchanges (DEXs)**: DEXs *"were born from the desire to address the vulnerabilities of centralized platforms"*,[34] so they aim to provide an exchange between one or more crypto-assets with non-custodial solutions. As a result, users are under complete control of their assets and do not need to transfer and store crypto-assets on an exchange. Based on the mechanism for price discovery, there may be different variants, such as:[35]

X. **Order Book DEX**: maintaining the state of an order book is a computationally expensive task, and, given the design of trustless and public distributed ledger technologies, it is not feasible to host it on-chain. Consequently, a user wanting to execute an order will sign a transaction allowing the DEX to execute the trade on his behalf when certain conditions are met. Orders are matched manually or

---

[32] https://github.com/aave/aave-protocol/blob/master/docs/Aave_Protocol_Whitepaper_v1_0.pdf. Accessed 3 May 2023.
[33] Perez et al. 2020, p 4.
[34] ESMA 2019, p 44.
[35] Werner et al. 2021, pp 3-4.

algorithmically; however, this will always involve a degree of trust in off-chain mechanisms[36] that may be susceptible to manipulation.[37]

XI. **Automated Market Maker (AMM)**: AMM is a decentralized protocol that relies on a mathematical formula to price assets. The AMM does not require an order book since the assets are priced according to a pricing algorithm. The liquidity to create the market is provided by the liquidity providers.

The liquidity providers provide funds to a liquidity pool, where reserves for two or more assets are locked into a smart contract. Each liquidity provider receives newly minted liquidity tokens representing their share of a pool's liquidity. A trade may occur when a user deposits an asset, thus providing more liquidity for that particular asset, and withdraws the reserves of one or more other tokens deposited in the pool. As the reserve ratios for a pool's assets change as liquidity is withdrawn and added, a liquidity provider may receive a different token ratio upon his liquidity share than the ratio he initially deposited. This risk is known as impermanent loss.[38]

XII. **Derivatives:** Derivatives are contracts that derive their value from the performance of an underlying asset, event, or outcome. Without an underlying to derive its value from, derivatives per se are valueless.

Since the development of derivatives contracts to help reduce the risk for farmers, the uses and types of derivatives contracts and the derivatives market size have increased significantly. Derivatives can be used to manage risks associated with the underlying, but they may also increase risk exposure for the other party to the contract. Nowadays, derivatives are no longer just used to reduce risk but are also part of the investment strategies of many fund managers and retail investors.

XIII. **Typical Derivatives**: Typical derivatives contracts, where futures and options are included, can be defined by having an underlying size and price, an expiration date, and a settlement mechanism.

XIV. **Perpetual Swaps**: These contracts are a mix of future contracts and contracts-for-difference, where there is no set expiry date, and the settlement may be satisfied via the delivery of the underlying asset. Furthermore, traders may decide to keep their position by providing a funding transaction if their position is underfunded.[39]

XV. **Synthetic Assets**: are collateral-backed tokens whose value fluctuates depending on the underlying asset reference index without directly taking a position in that asset. Some platforms allow the creation of crypto-backed synthetic assets that provide on-chain price exposure to commodities, stocks, indices, or other assets.[40]

---

[36] https://idex.io/document/IDEX-2-0-Whitepaper-2019-10-31.pdf. Accessed 4 May 2023.
[37] Daian et al. 2019, p 1.
[38] https://uniswap.org/docs/v2/advanced-topics/understanding-returns/. Accessed 4 May 2023.
[39] https://www.bitmex.com/app/perpetualContractsGuide. Accessed 4 May 2021.
[40] https://synthetix.io/. Accessed 4 May 2023.

XVI. **Portfolio Management**: DeFi protocols allow the automation of on-chain assets portfolio management. Crypto-assets are deposited and pooled into smart contracts and allocated according to an investment strategy encoded in the smart contract or managed by a pool manager. The investment strategy may vary from rebalancing the portfolio[41] to yield aggregating strategies.[42]

XVII. **Payments**: Blockchain allows anyone to send value without a trusted intermediary or depository; however, a transaction may take some time to be confirmed and irreversible in the blockchain. DeFi payment protocols allow for processing instant and high-volume micropayments that may remove the risk of delegating custody of the crypto-assets to trusted third parties.[43]

XVIII. **Insurance**: Insurance is a risk management strategy in which an individual receives financial protection or reimbursement against losses from an insurance company in an unfortunate incident. DeFi insurance protocols are, so far, restricted to on-chain risk coverage, where any agent may purchase cover for unintended uses of code in any smart contract offered from the list of the insurance protocol. When an event similar to "The DAO" hack happens, the agent would be covered from any financial losses resulting from such hacking.[44]

6.4 How Does the DLT Pilot Regime Match with DeFi?

One of the purposes of the DLT Pilot Regime is to take advantage of the potential of DLT, a transformative technology in the financial sector, in reducing intermediaries and, subsequently, transaction costs. This purpose is related to multiple benefits, such as enhanced efficiency, transparency, and competition in relation to trading and settlement activities.

In this context, many repetitive business processes, namely the input of transaction data separately in each layer of the custody chain, together with costly reconciliation, can be eliminated.[45] The prominent example of the use of this potential of DLT is what is allowed to a DLT TSS, as presented in Fig. 6.2.

The DLT Pilot Regime allows an investment firm or market operator under MiFID II to operate a DLT TSS and allows the same exemptions as those available to operators of DLT MTFs and of DLT SSs, including the obligation of intermediation in the access of retail investors to the market infrastructure as a member or participant.

Figure 6.2 represents the traditional operation of trading and settlement services. What a DLT TSS can do to that traditional operation under the DLT Pilot Regime is eliminate the need for brokers and a CSD to access the market, custody and settlement

---

[41] https://www.dhedge.org/. Accessed 4 May 2023.
[42] https://yearn.finance. Accessed 4 May 2023.
[43] https://matic.network/. Accessed 4 May 2023.
[44] https://nexusmutual.io/. Accessed 4 May 2023.
[45] Priem 2022, p 374.

(represented by the crosses in Fig. 6.2). With a DLT TSS, only one operator is needed in the value chain of a financial instruments market.

However, it is important to consider that in some DeFi applications, no natural or legal person performs or provides trading or settlement services.[46] Those applications are considered "decentralized", but this term is used in several different senses across the DLT industry.

Firstly, the term is used in the Settlement Layer, where a network of nodes comprises a permissionless blockchain through peer-to-peer connections between unrelated and independent agents, rather than relying on a central server or a central organization. This axis of decentralization is referred to as the architectural distribution.[47] For the purpose of this chapter, we shall assume that the Settlement Layer in which the DeFi protocols are built is always sufficiently distributed.[48]

Secondly, the term may refer to the decentralization of crypto-assets custody (non-custodial crypto-assets by an intermediary). In traditional finance, custody is a core financial service in which the custodian holders in safekeeping assets on behalf of the client. Usually, this service is endorsed by the custodian's reputation and the legal and regulatory framework. The core reason for using custodians is the security of not having the assets lost or stolen. As a result, security and trustworthiness become symbiotic pertaining to the safekeeping of assets. With the technological forthcoming of DeFi applications, the solution for proving trustworthiness without endorsement from central authorities is via a robust, resistant, and transparent technical implementation of a set of functions, in which only the client/user has access to the assets deposited in a particular smart contract.[49] In the Application Layer, a non-custodial protocol allows users to have complete control of their crypto-assets deposited in a smart contract address for a specific finality of a protocol without relying on a centralized party. The interaction between the protocol and the user is autonomous and automatic, without any contact with the development team, thus, transferring the responsibility from the intermediaries to the users.

Thirdly, the term is referred to the decentralization of management, organization, and ownership of a protocol. Even though this type of decentralization is fundamental across all Layers, there is a clear trend of "decentralized governance" across DeFi protocols. Organizations that seek to emulate the operation of a corporate entity through code and integrate decentralized governance features within their protocols are usually known as Decentralized Autonomous Organizations (DAOs),[50] where the stakeholders hold voting rights via tokens that determine an autonomous execution of the nexus of smart contracts when certain conditions are met. The tokens

---

[46] Schillig 2021, p 37.
[47] https://medium.com/@VitalikButerin/the-meaning-of-decentralization-a0c92b76a274. Accessed 5 May 2023.
[48] For further details pertaining the "decentralization" discussion of the Settlement Layer, see: Walch 2019.
[49] ESMA 2019, p 44.
[50] "A DAO may be defined as a smart contract conceptualized as a relatively autonomous and self-sufficient for-profit organization, which is jointly held by token holders and may share its earnings." See Rolo 2019, p 59.

representing the possibility for token holders to be part of a DeFi protocol are known as governance tokens and may give the token holder control (voting rights)[51] and ownership (cashflow rights) of the organization.

Since DeFi is not composed of watertight compartments,[52] the requirement for a central and liable party to be the operator of a DLT market infrastructure may thus be difficult to achieve. Therefore, some degree of centralization in DeFi applications may be needed and encouraged to respect the rule of law and prevent risks and efficiency.

Intermediation is fundamental for finance operations, as commercial banks, investment banks, stockbrokers, mutual funds, insurance companies and stock exchanges form the fabric of all frameworks of modern financial structures.[53] As a functionally interconnected institutional structure is inherent to the financial system, the financial regulatory frameworks are always based on an *ex-ante* intervention (authorization procedures) and the creation of rules of conduct for intermediaries.

As noted by Organization for Economic Cooperation and Development (OECD) and the Financial Stability Board (FSB), most DeFi projects often exist along a spectrum of centralization, depending on the stage of development of the application,[54] which generally means having an identifiable intermediary that would be the liable entity within DLT Pilot Regime.

Additionally, as stated in the latest recommendations of the Financial Action Task Force (FATF), DeFi *"applications or platforms are often run on a distributed ledger but still usually have a central party with some measure of involvement, such as creating and launching an asset, setting parameters, holding an administrative " key" or collecting fees"*.[55]

Lastly, DeFi applications are likely to evolve into increasingly centralized platforms with the market pressure, i.e., the emergence of new intermediary operators and new potential incumbents in the DeFi space.[56]

In conclusion, DLT Pilot Regime can cover some DeFi categories, such as DEXs and stablecoins. Naturally, it will be impossible to have "true decentralized" applications within the DLT Pilot Regime because this Regulation requires an operator of a DLT market infrastructure to be a liable party in the provision of the services. Even so, this Regulation will be an important step for some DeFi' objectives, particularly disintermediation and fostering non-custodial business models.

---

[51] "Instead of asset ownership, governance token ownership applies to voting rights, as the name suggests", Harvey et al. 2021, p 21.

[52] Zunzunegui 2022, p 11.

[53] Lin 2015, p 1.

[54] OECD 2022, p. 20; FSB 2022, p 16.

[55] FAFT 2020, p 23.

[56] De Filippi and Lavayssière 2020, p 204.

References

Correia F (2017) A tecnologia descentralizada de registo de dados (*Blockchain*) no sector financeiro. In: Cordeiro AM, Perestrelo de Oliveira A, Pereira Duarte D (eds) FinTech. Desafios da Tecnologia Financeira. Almedina, Coimbra, pp 69-74

Daian P, Goldfeder S, Kell T, Li Y, Zhao X, Bentov I, Breidenbach L, Juels A (2019) Flash boys 2.0: Frontrunning, transaction reordering, and consensus instability in decentralised exchanges. Available at: https://arxiv.org/abs/1904.05234. Accessed 4 May 2021

De Filippi P, Lavayssière X (2020) Blockchain Technology: Toward a Decentralised Governance of Digital Platforms? Available at: https://hal.archives-ouvertes.fr/hal-03098502/document. Accessed 10 March 2021

De Filippi P, Wright A (2018) Blockchain and the Law: The Rule of Code. Harvard University Press

Finlow-Bates K (2020) Move over brokers here comes the blockchain. Thinklair TMI

FSB (2022) Assessment of Risks to Financial Stability from Crypto-assets. Available at: https://www.fsb.org/wp-content/uploads/P160222.pdf. Accessed 10 March 2022

Harvey C, Ramachandran A, Santoro J (2021) DeFi and the Future of Finance. Available at: https://papers.ssrn.com/sol3/papers.cfm?abstract_id=3711777. Accessed 3 May 2021

Katona T (2021) Decentralized Finance - The Possibilities of a Blockchain "Money Lego" System. Financial and Economic Review, Magyar Nemzeti Bank (Central Bank of Hungary), vol. 20(1), pp 74–102

Klages-Mundt A, Harz D, Gudgeon L, Liu J, Minca A (2020) Stablecoins 2.0: Economic foundations and risk-based models. Proceedings of the 2nd ACM Conference on Advances in Financial Technologies

Lau D, Lau D, Sze Jin T, Kho K, Azmi E, Lee T, Ong B (2020) How to Defi. CoinGecko

Lin T (2015) Infinite Financial Intermediation. Wake Forest Law Review, Vol. 50, No. 643, Temple University Legal Studies Research Paper No. 2016–06

Maia G, Santos J (2022) MiCA and DeFi ("Proposal for a Regulation on Market in Crypto-assets" and "Decentralised Finance"). Revista Electrónica de Direito, n.º 2, Volume 28, 2022, pp. 58–82. Available at: https://cije.up.pt/pt/red/ultima-edicao/mica-e-defi-ldquoproposta-de-regulamento-sobre-mercados-de-criptoativosrdquo-e-financas-descentralizadasrdquo/ Accessed 25 July 2022

Perez D, Werner S, Xu J, Livshits B (2020) Liquidations: Defi on a knife-edge. Available at: arXiv: 2009.13235. Accessed 3 May 2021

Perumal A, Deepica G, Bindu V (2018) Simulation of Compiler Phases. Indo-Iranian Journal of Scientific Research (IIJSR), Peer-Reviewed Quarterly International Journal, Volume 2, Issue 2, pp 100–104

Priem R (2022) A European DLT pilot regime for market infrastructures: Finding a balance between innovation, investor protection, and financial stability. Journal of Financial Regulation and Compliance; 2022; Vol. 30; Iss. 3; pp. 371–390

Raskin M (2017) The Law and Legality of Smart Contracts, 1 Geo. L. Tec. Rev. p. 306

Rolo A (2019) Challenges in the Legal Qualification of Decentralised Autonomous Organisations (DAOs): The Rise of the Crypto-Partnership? Revista de Direito e Tecnologia, Vol 1, no. 1. pp 33-87

Schär F (2021) Decentralized Finance: On Blockchain and Smart Contract-based Financial Markets. Federal Reserve Bank of St. Louis Research Paper Series

Schillig M (2021) Lex Cryptographia, 'Cloud Crypto Land' or What?—Blockchain Technology on the Legal Hype Cycle. Available at: https://papers.ssrn.com/sol3/papers.cfm?abstract_id=3804197. Accessed 3 May 2021

Walch A (2019) Deconstructing "Decentralization": Exploring the Core Claim of Crypto Systems. In: Brummer C (ed) Cryptoassets: Legal, Regulatory, and Monetary Perspectives. Oxford University Press, Oxford. Available at https://academic.oup.com/book/35207

Werner S, Perez D, Gudgeon L, Klages-Mundt A, Harz D, Knottenbelt W (2021) SoK: Decentralized Finance (DeFi). Available at: arXiv:2101.08778. Accessed 3 May 2021

Zetzsche D, Annunziata F, Arner D, Buckley R (2020a) The Markets in Crypto-Assets Regulation (MICA) and the EU Digital Finance Strategy. European Banking Institute Working Paper Series, No. 2020/77, available at: https://papers.ssrn.com/sol3/papers.cfm?abstract_id=3725395 Accessed 25 July 2022

Zetzsche D, Arner D, Buckley R (2020b) Decentralized Finance (DeFi), Journal of Financial Regulation, 2020, 6, pp 172–203, Available at: SSRN: https://ssrn.com/abstract=3539194 or https://doi.org/10.2139/ssrn.3539194. (last revised 11 November 2021) Accessed 25 May 2023

Zetzsche D, Woxholth J (2021), The DLT Sandbox under the Pilot-Regulation. European Banking Institute Working Paper Series No. 2021/92. Available at: The DLT Sandbox Under the EU Pilot Regulation, Oxford Law Faculty. Accessed 12 February 2022

Zunzunegui F (2022) How to regulate digital financial platforms: A research agenda. Revista de Derecho del Mercado Financiero, Working Paper 3/2022, available at: http://www.rdmf.es/wp-content/uploads/2022/10/Zunzunegui.-F.-How-to-regulate-digital-financial-platforms-A-research-agenda.pdf

Other Documents

ESMA (2019) Advice Initial Coin Offerings and Crypto-Assets. https://www.esma.europa.eu/sites/default/files/library/esma50-157-1391_crypto_advice.pdf. Accessed 5 May 2021

FATF (2020) Draft updated Guidance for a risk-based approach to virtual assets and VASPs. http://www.fatf-gafi.org/media/fatf/documents/recommendations/March%202021%20-%20VA%20Guidance%20update%20-%20Sixth%20draft%20-%20Public%20consultation.pdf Accessed 28 April 2021

OECD (2022) Why Decentralised Finance (DeFi) Matters and the Policy Implications. Available at: https://www.oecd.org/daf/fin/financial-markets/Why-Decentralised-Finance-DeFi-Matters-and-the-Policy-Implications.pdf. Accessed 10 March 2022

Chapter 7
Central Bank Digital Currency: A Focus on Anonymity

Anjeza Beja and Bernardo Correia Barradas

Contents

7.1	Introduction	108
7.2	Anonymity Options and Author's Survey Results	110
	7.2.1 But How Much Do Consumers Value Anonymity in Payments and in CBDC if Issued by a Central Bank for General Use?	111
7.3	Technological Options	113
7.4	Major Legal Implications of CBDC Implementation	116
7.5	Difference Between Central Bank Money and Private Money—Legal Mandate	116
7.6	Legal Tender	120
7.7	El Salvador Case	121
	7.7.1 Country Context	121
	7.7.2 Bitcoin as Legal Tender	122
References		124

Abstract Blockchain has been uppermost in the news almost uninterruptedly since it entered the mainstream with the appearance of several different crypto-assets (This term includes cryptocurrencies, virtual assets, virtual currencies, digital assets, digital currencies, and similar definitions used worldwide, and encompasses "tokens" used or classified as payment tokens, security tokens, utility tokens, and stablecoins, among others.) and its availability to the general public. The interest of the authorities (in particular during the last half of 2022) has been growing exponentially and is leading to a renewed focus on what was, until now, known as the Crypto "Wild West" (https://www.wsj.com/articles/sec-will-police-cryptocurrencies-to-maximum-possible-extent-chair-gary-gensler-says-11628007567.) Further to the legal and regulatory considerations arising from the crypto ecosystem, authorities have also been looking into the possibilities that can be harnessed from its underlying technologies and concepts, and testing solutions such as those related to land registry, social-benefits management, patent protection and digital currencies

---

A. Beja
Prokop Myzegari, 1016 Tirana, Albania
e-mail: anjezaharizi@gmail.com

B. Correia Barradas (✉)
13090 Aix-en-Provence, France
e-mail: bernardo@correiabarradas.com

issued by central banks (Central Bank Digital Currencies or CBDCs). To be noted, however, is that all those activities, although they could use blockchain or DLT-based solutions, do not necessarily need to. This includes CBDC, which can be issued and managed through blockchain-type solutions, traditional methods, or a combination thereof. In this chapter we consider the topic of CBDC, what it is and its implications with a particular emphasis on anonymity—an aspect that we believe will be of the utmost importance within the implementation of CBDC by central banks around the world, in particular when, and if, central banks from major and advanced economies move forward with CBDC solutions.

Keywords Blockchain · Central Bank · Digital currency · CBDC · Crypto

7.1 Introduction

CBDC is the topic of the day. Central banks' interest and public interventions on CBDCs have turned more positive since late 2018, with a rapid increase starting in the second half of 2020 and the first quarter of 2021.[1] Although Covid-19 brought many challenges for central banks and governments around the world, some opportunities were also identified, such as a decline in the use of cash for payments and consumers shifting from physical to digital payments, thus adding new motivations and opportunities for possible CBDC implementation.

The widely accepted definition is that *CBDC is central bank-issued digital money denominated in the national unit of account, and it represents a liability of the central bank*. CBDC can be "retail" (i.e. for general purpose), or "wholesale", for restricted access by financial institutions. The main prominent difference between CBDC and cash, is the form. CBDC comes in a digital form, unlike cash which comes in physical coins and banknotes.

According to the results of the third Bank for International Settlements (BIS) survey on central bank digital currency,[2] there are several motivations or policy goals for central banks to consider issuing a retail CBDC. These include financial stability, monetary policy implementation and financial inclusion as well as payments efficiency and safety. Not all jurisdictions give the same weight to each of these motivations, which is related to their economic environment and characteristics of the market.[3] For example, developing economies prioritize financial inclusion and consider the implementation of CBDC as a possible solution for this aim. Nevertheless, domestic payments efficiency and payments safety remain at the heart of both advanced and developing economies' motivations for issuing general purpose CBDC. Preserving financial stability and supporting monetary policy, although

---

[1] Auer et al. 2020.

[2] https://www.bis.org/publ/bppdf/bispap125.htm.

[3] For example, in more advanced economies with modern and inclusive payment systems, the need for a CBDC moves away from the goal of financial inclusion.

not top priority motivations, are being considered more extensively by developing economies compared to the opposite trend in advanced economies.[4] According to the same survey, regarding the wholesale issuance of CBDC, the motivations tend to be similar to retail CBDC except for cross-border payments efficiency, where motivations for wholesale CBDCs are more accentuated. Other important motivations include development of capital markets, enhancement of cyber resilience, and improvements in securities trading and settlement.

Beside the above, the BIS survey also pointed out some implicit motivations making central banks consider CBDC, such as the preservation of the monetary sovereignty of one country's currency.[5] This includes two aspects: first, a local currency CBDC could enhance the usage of the local currency compared to another widely used and accepted foreign currency (such as USD in dollarized economies or EUR in euroized economies), given that the public could be more attracted to a digital form of the national currency rather than to a materialized form of a foreign currency; and secondly, a CBDC could address the competition concerns on the local currency due to the increased use of private cryptocurrencies or stablecoins such as Tether (USDT) and USD Coin (USDC).

The abovementioned motivations or policy goals were confirmed by a more recent study published by the International Monetary Fund (IMF).[6] The study included six countries in different positions with regard to their progress in CBDC implementation, including central banks which have already started issuing CBDC, central banks in a piloting phase as well as central banks which have decided against issuing CBDC for the time being.

Despite the different motivations and potential benefits, CBDC implementation is associated with several implications which need to be carefully considered by central banks. These include the legal implications of CBDC implementation such as legal tender adoption, as well as cyber risks, data privacy, accounting implications and financial stability. On the other hand, technological alternatives, still emerging, require careful assessment to appropriately address and fulfil the policy goals mentioned above.

Another relevant aspect related to the design features of CBDC, that is still largely under discussion among the different authorities, is the anonymity as to the ownership of, and transactions in, digital currencies issued by central banks. In terms of ownership, currently banknotes are largely anonymous, whilst transactions, above a certain limit, on bank or payment accounts can be traced due to national AML/CFT (anti-money laundering and countering the financing of terrorism) requirements.

But is it legally possible to have a central bank digital currency fully or at least, partially anonymous? Or will central banks and other authorities use this opportunity to finally fully track money used by corporations and individuals and address AML/

---

[4] Boar and Wehrli 2021.

[5] See for example the references to monetary sovereignty in European Central Bank (2020) Report on Digital Euro. https://www.ecb.europa.eu/paym/digital_euro/report/html/index.en.html. Accessed 23 May 2023.

[6] Behind the Scenes of Central Bank Digital Currency, Fintech notes, IMF, 2022.

CFT risks? How will the public react to all this? And what about the technology? Are central banks more in favor of using centralized technology based on centralized ledgers? Or are they favoring the use of decentralized ledger technologies, such as blockchain, as a more suitable option for implementing CBDC?

Since the topics above may become a significant issue when it comes to the public's willingness to adopt and use digital currencies, and also when it comes central banks' constraints on issuing CBDC, we have decided to analyze in this chapter the abovementioned questions, respectively based on the legal framework in the European context and based on some central banks' decisions on the technology choices.

7.2 Anonymity Options and Author's Survey Results

In October 2020, seven central banks together with the BIS published a report[7] identifying the key principles necessary for any publicly available CBDCs. Although the central banks that contributed to this report have not yet decided whether or not to issue a CBDC, their common motivation to engage in such a project was to use CBDC as a means of payment. They outlined the following set of key principles which should be considered when issuing CBDCs:

- coexistence with cash and other types of money in a flexible and innovative payment system;
- any introduction should support wider policy objectives and do no harm to monetary and financial stability;
- features should promote innovation and efficiency.

Considering that the main feature of cash is that there is no record of either identity or the transactions themselves, it won't be that easy from a customer's point of view moving into a fully traceable means of payment (aside from other features that may, from a technical point of view, be implemented with CBDC such as limits or restrictions on certain type of payments, reversibility, and application of negative interest rates). Nevertheless, full anonymity is not considered a plausible option in the abovementioned report. Although addressing AML/CFT concerns is not listed among the primary motivations from a central bank perspective, the latter is expected to comply with these rules when issuing CBDC to the public.

According to a paper issued by Bech and Garratt (2017)[8] there could be two anonymity options: (i) counterparty anonymity where the payer does not need to reveal its true identity to the recipient of funds and (ii) third-party anonymity where the payer does not reveal its true identity to other members of the community (e.g., a payment service provider, bitcoin or other cryptocurrency community or central

---

[7] Central bank digital currencies: foundational principles and core features, https://www.bis.org/publ/bppdf/bispap125.htm.

[8] Bech and Garratt 2017.

bank in the case of the CBDC). The authors state that counterparty anonymity seems less controversial than third-party anonymity. The latter should not be allowed as it may facilitate criminal activity, such as tax evasion, terrorist financing or money laundering. Considering that from a technological standpoint there are no limitations, it seems like the approach followed will rely on the policy decision of the different central banks. For hundreds of years central banks or equivalent entities have issued anonymous cash and customers have not had direct access to its accounts. But, as Bech and Garratt state,[9] the anonymity properties of cash are likely to have emerged out of convenience or historical happenstance rather than intent of the authorities.[10]

The Bank of England (BoE) in its Discussion Paper published in 2020[11] debated the possibility of a partially anonymous CBDC, which could be designed to protect privacy and give users control over who they share data with, even if CBDC payments are not truly anonymous (or secret). For example, a user may legitimately want to make a payment to a supermarket without sharing their identity with the supermarket, as this would allow the merchant to build a picture of their shopping habits. Therefore, according to this option, it could be possible to have counterparty anonymity without having third parties' anonymity as required by law.

7.2.1 But How Much Do Consumers Value Anonymity in Payments and in CBDC if Issued by a Central Bank for General Use?

For the purposes of this chapter, we undertook a survey where 104 people participated. Our target group was mostly people who had either a solid theoretical knowledge or a basic one about CBDC, including central bankers/regulators (40%), private payment service providers (14%), development/international organizations (11%), lawyers/consultancy (13%), and academia (10%). The participants are from four continents, with a slight majority coming from Europe (55%). Three main questions were addressed to the participants, which are further detailed below.

To the question on "*how much do you value anonymity to protect your privacy when making payment transactions online*", most of the respondents (77%) answered that they do not value it that much since they generally trust the regulated and supervised payment service providers, although, they mostly prefer credit institutions' online payment applications to make their payments. Only 21% answered that they are very sensitive to anonymity and as a result they never make payment transactions online as they don't want to share their personal data with any payment platform.

---

[9] Ibid.

[10] On the other hand, the same can be said of digital payments, where traceability or the possibility of it -for example in our credit and debit card transactions- emerged out of circumstances and not necessarily by policy or intent.

[11] Bank of England 2020.

To the question of *"what type of anonymity would you prefer (counterparty versus third party)"*, if that was a choice put to them, most of the respondents (around 39%) said they would prefer to have full anonymity. For those who choose between the two types, only 10% preferred counterparty anonymity and 30% third party anonymity. Within this category, it appears that people are more concerned about the type of information that payment service providers collect and administrate compared to the information collected by the recipients of funds. 21% of respondents were indifferent to any of the options.

The final question was about the respondents' concerns about anonymity if the central bank itself were to issue a CBDC. Considering that a significant part of the respondents came from central banks, the results to this answer were not unsurprising. The most preferred answer (58%) was that they are aware of the risks associated with anonymity in payment transactions (e.g., AML/CFT) and therefore they would expect that anonymity be applied *only* for payments below a certain threshold. This seems to be the same approach that different jurisdictions are applying for cash payments for anti-money laundering and counter terrorism financing purposes. Still, there were 26% of participants who said that they would expect the CBDC to be fully anonymous, same as cash so that the central bank won't be able to track or monitor their transactions or spending habits. This result is in line with the result of the first question, where 21% answered they were concerned about privacy and didn't want to share their personal data with any payment platform. 16% of the respondents fully embraced the possibility of a non-anonymous CBDC because they consider it as a good opportunity for central banks to address informality in the economy. As a result of this survey, we extracted the following outcomes:

> Firstly: anonymity does not seem to be a general concern for our audience because sharing their true identity with regulated payment service providers is not a novelty. They are already used to sharing personal data with payment service providers, although most of them prefer credit institutions.[12] Strong data protection legislation and enforcement and customer protection rules are required from regulators to fortify this perception with the introduction of the CBDC. On the other hand, payment service providers should efficiently mitigate cyber risks to complete the picture and maintain customer trust in the future.

> Secondly: if people were to choose, they would prefer making completely anonymous transactions, but on the other hand they understand the importance of not doing so for the economy as a whole. Therefore, the AML/CFT requirements currently applied for cash payments, which require identification for payments above a certain threshold, are endorsed by them as a solution and are expected to be applied as well for CBDCs. This might be trickier for central banks which are planning to issue directly CBDC to the public, as they will have to establish their own monitoring structures and capacities to ensure this.

Based on the comprehensive analysis of the European Central Bank (ECB)[13] public consultation on the Report on digital Euro (with a record number of respondents), what the public and professionals want the most from such a digital currency is

---

[12] In all of the surveyed cases credit institutions were banks.

[13] A comprehensive list of documents on recent developments of the digital euro can be found here: https://www.ecb.europa.eu/home/search/html/digital_euro.en.html.

privacy (43% of respondents).[14] Nevertheless, neither group showed support for full anonymity, but supported requirements to avoid illicit activities. The results of this report are in line with the results in our survey conducted for the preparation of this chapter.

According to the proof of concept[15] that has been developed by the European System of Central Banks (ESCB) EURO chain research network, from a technological standpoint, it is possible to make use of anonymity vouchers which allows users to process low-value transactions without revealing their identities. Distributed Ledger Technologies (DLT) can be used to balance an individual's right to privacy (for lower value transactions) with the public's interest in the enforcement of AML/CFT regulations (for higher-value transactions).

The European Commission, through the new proposed regulation on AML/CFT,[16] is planning to ban cash transactions of more than EUR 10 000 for AML purposes which could be interpreted as a steppingstone towards the preparation of the public to go cashless and embrace a digital euro more readily, should it be introduced in the future.[17]

In Table 7.1, we summarize the pros and cons of the different levels of anonymity in a CBDC design discussed above.

7.3 Technological Options[18]

While a fast pace in technological development can be a positive factor, this is not helping central banks in their decision-making process in designing CBDCs. Typically, central banks operate in a centralized ledger system, while DLT solutions, of which blockchain is arguably the best known, brought to the fore a new technological choice. Since DLT is still in a developing phase, it brings a lot of uncertainties related to its capacity and suitability. That being said, there is not a universal approach among central banks to embrace DLT as the main technology enabling the issuance of CBDC.

Central banks which are in advanced phases of CBDC testing or piloting are assessing DLT application versus centralized ledgers to identify constraints and limitations. The People's Bank of China (PBC) which has already tested DLT in its pilots, decided that it has limited capacity to process transactions, especially on days with high levels of transactions. For this reason, the PBC has committed to a "hybrid architecture", meaning that the DLT will be used in limited areas, allowing also for

---

[14] https://www.ecb.europa.eu/press/pr/date/2021/html/ecb.pr210414~ca3013c852.fr.html.

[15] https://www.ecb.europa.eu/paym/intro/publications/pdf/ecb.mipinfocus191217.en.pdf.

[16] https://finance.ec.europa.eu/publications/anti-money-laundering-and-countering-financing-terrorism-legislative-package_en#regulation.

[17] On limitation on the use of cash please see CJEU's ruling on joined Cases C-422/19 and C-423/19.

[18] Behind the Scenes of Central Bank Digital Currency, Fintech notes, IMF, 2022.

Table 7.1 Pros and cons of the different levels of anonymity in a CBDC design

	Full anonymity	Partial anonymity	No anonymity
Pros	The CBDC will be cash-like, except for the form. People will embrace it more easily since their perception of using cash will not be different and they will continue making payments in cash as they have been doing. According to Kahn et al (2005) and Mc Andrews (2017), payees and payers might support this option more because they may want to reduce the risk of identity theft, the possibility that the counterparty might follow them home and assault them, or more innocuous annoyances like directed advertising and solicitations (spamming). The faster people embrace the new digital currency, the easier it will be implemented and successfully distributed.	Partial anonymity enables anonymous transactions for lower amounts and/or for a limited number of transactions within a certain time. While for high value transactions or above a certain threshold, they will have to be identified. This option enables people to perform high value transactions in a secure way while also providing flexibility for the execution of low value payments.	Ensures integrity of the financial transactions Ensures full coverage of funds or protection if money is lost or stolen.

(continued)

Table 7.1 (continued)

	Full anonymity	Partial anonymity	No anonymity
Cons	Increases risk of ML/FT since it could be used for criminal activities; no security and protection if the digital euro is lost since the holder cannot be identified or prove that the lost money belonged to him or her. This is particularly problematic for economies relying on cash usage and with low financial inclusion rates since trust in the currency could be questioned.	Requires close institutional cooperation and coordination to ensure that the established limits for being anonymous or not are duly respected. Also reducing the amount of information visible to parties not involved in the transaction remains a challenge.[19]	People and transactions will be traceable and identified constantly. Hence individuals' privacy in doing transactions is monitored. This could create opportunity for more surveillance and censorship from Governments (e.g., countries with a weak rule of law, incapable institutions, and little political accountability).[20] There is no assurance that the customers' data will not be misused either for political reasons or subject to cyber-attacks. Since close monitoring will be required to the central bank for AML purposes, which is an activity it has not been involved in the past directly, reputational risks might also arise.

Source The authors

---

[19] European Central Bank (2020) Report on Digital Euro. https://www.ecb.europa.eu/paym/digital_euro/report/html/index.en.html. Accessed 23 May 2023.

[20] For example, authorities could in theory block payments to companies or businesses that they wish to punish or collect and use people's transactional data for electoral purposes or to induce specific habits.

other technologies to be featured on its core centralized ledger. A similar approach could be followed by the Central Bank of Uruguay in the second pilot of the e-peso. The first pilot was not based on DLT.

The Central Bank of the Bahamas (CBOB)—the first central bank in the world to issue a CBDC (Sand Dollar) and the Central Bank of Nigeria CBCD (e-Naira) are both deploying DLT.[21] The Bank of Jamaica which also launched its CBDC Jam-Dex in 2022, does not deploy DLT, but instead uses traditional centralized systems.[22] The Eastern Caribbean Central Bank (ECCB) which is currently in a 2-year pilot phase, is deploying blockchain technology. The Sveriges Riksbank is currently exploring a DLT-based proof of concept, but the chances are that their potential future e-krona might possibly be built with reference to a different and more conventional technology. In Fig. 7.1, we present a summarized overview of the current CBDC implementation trends and use cases by some central banks.

7.4 Major Legal Implications of CBDC Implementation

The more central banks advance with their CBDC implementation projects, the more questions and doubts arise related to legal and regulatory implications. This is due to the fact that CBDC is not just a technical project, whose successful implementation relies solely on technological issues. Beside the policy and political issues which may arise, and which vary from jurisdiction to jurisdiction, a careful assessment of legislative aspects should be undertaken. Central banks laws and monetary laws are at the forefront in this regard, but other primary legislation such as commercial banks laws, tax laws, private laws, payment systems and payment services laws, data protection laws, AML/CFT laws etc. could also be subject to evaluation and probably revision.[23] Below in the following sections we provide a quick overview of two probable main legal implications of CBDC implementation: the difference between central bank money and private money, and the legal tender characteristic of central bank money and how such would be applied to CBDC.

7.5 Difference Between Central Bank Money and Private Money—Legal Mandate

Making use of the definition mentioned in Introduction above, CBDC represents a liability of the central bank. Comparable to cash in circulation or to commercial bank accounts with central banks which represent a liability account at the central

---

[21] https://www.enaira.gov.ng/about/faq.
[22] https://www.ledgerinsights.com/jamaican-central-bank-plans-digital-currency-but-not-blockchain/.
[23] Bossu et al. 2020.

7 Central Bank Digital Currency: A Focus on Anonymity

According to a recent survey performed by the BIS[a], which included 65 central banks from around the globe, central banks interest in CBDC is rising further. In 2020, about 60% of central banks (up from 42% in 2019) were conducting experiments or proofs-of-concept, while 14% were moving forward to development and pilot arrangements. The central banks not currently involved in any CBDC work are primarily in smaller jurisdictions. We provide below a review of the most important current use cases in advanced stages as well as the respective approaches when it comes to CBDC anonymity and technological choices.

- The first CBDC, Sand Dollar, introduced by the CBOB went live on October 21, 2020. Some of the benefits of introducing a digital currency for the Bahamian economy include a potential suppression of economic costs associated with cash usage, increase in the financial inclusion level of the population and benefits to the Government from improved expenditure and tax administration systems. The digital version of the Bahamian dollar is available for both wholesale and retail applications. One of the aspects, linked also to the topic of this chapter, is that the Sand Dollar is not anonymous, and it includes a fully auditable transactions trail. Transactions monitoring still protects user confidentiality and is governed by strict regulatory standards around access. Beside the currency issuance, the central bank is sponsoring a centralized KYC/identity infrastructure[b]. The CBDC issuing system is DLT-based.
- The second CBDC, Dcash, introduced by the ECCB, started a twelve-month pilot on March 31, 2021, on four of the eight island countries under its currency union. The main objectives for issuing Dcash include increasing opportunities for financial inclusion, growth, competitiveness, and resilience for citizens of the ECCU (Eastern Caribbean Currency Union). The digital version of the EC Dollar is available for financial transactions between consumers and merchants, people-to-people (P2P) transactions, all using smart devices. If a user does not have a bank account, that user can have a value-based wallet of Dcash with lower barriers to entry and correspondingly lower transactional limits. This wallet intends to address the goal of financial inclusion. According to the information published on the ECCB website, during the pilot phase users can benefit from privacy and confidentiality of their transactions. Nevertheless, since the registration is managed by financial institutions, they must handle all CDD requirements etc. Transaction and holding limits based on AML rules are applied. Dcash will be accessible even for those not having a bank account through value-based wallets, where lower entry barrier and limits apply[c]. The CBDC piloting system is DLT-based.
- The third CBDC project, e-Peso, introduced by the Central Bank of Uruguay (CBU), started a pilot on November 3, 2017, which ended six months later. The main motivation for starting this project was the reduction of transaction costs in payments, especially due to the cost of producing and transporting physical cash. E-Peso transactions are made anonymously on the "virtual vault" component of the system. This is a shared feature with paper money. Second, and differently from paper money, e-Peso transactions can be fully traced. Although the virtual vault stores transactional data per (anonymous) digital wallet, this data could be decrypted in order to reveal the identity of the user making transactions through a particular wallet for legal purposes (e.g., a judge of law, a tax authority, or an anti-money laundry authority)[d]. The CBDC piloting system at the CBU is not DLT-based.
- The Sveriges Riksbank (SR) is also in an advanced phase with its e-Krona project. The Swedish central bank announced recently that it is moving from having simulated participants to cooperation with external participants in its test environment. The main motivation for starting this project is the decline in the usage of cash in Sweden[e]. The SR sees potential problems arising from this and is therefore running a project to investigate the possibility of producing a digital complement to cash. The report published in April 2021 provides more information on the technical solution tested (token-based in a distributed network based on blockchain technology) and the legal analyses of the solution (compared to the means of payment existing today and AML regulations), but it specifies, though, that no decision has been taken yet by the SR on whether to issue an e-krona or on how the e-krona would be designed and which technology would be used. Regarding anonymity, accounts may not be anonymous pursuant to current legislation in Sweden, although it is possible to perform anonymous transactions on very limited use cases[f]. The SR is currently exploring a DLT-based proof of concept, but in their view a potential future e-krona does not necessarily have to be built on DLT. A second e-krona proof of concept or pilot could thus be based on a different technology.
- The PBC is at the forefront of CBDC development with the e-yuan. Starting October 2020, it has launched the first pilot of the DC/EP (Digital Currency/Electronic Payment) or e-CNY project. By April

Fig. 7.1 Current CBDC implementation trends and use case review.[24] *Source* The authors

---

[24] The references related to the technological choices for CBOB, ECCB, CBU, SR, POBC are from Behind the Scenes of CBDC, Fintech notes, IMF, 2022.

2021 seven rounds of digital renminbi red envelopes have been piloted, with a total of 150 million yuan distributed, and more than 750,000 people received and used it. The PBC did not rely on its own efforts to promote the digital renminbi, but extensively combined with traditional financial institutions, internet companies, and retail merchants to integrate the issuance, circulation, and transaction scenarios.[g] In 2022 the PBC has expanded its prototype central bank digital currency (CBDC), allowing it to be used on buses and for housing payments.[21] Furthermore, the PBC has advanced in enabling the use of e-yuan for cross border payments. It is one of four central banks being part of the m-Bridge project launched by the BIS Innovation hub in Hong Kong.[i] Regarding anonymity, according to the director of the Digital Currency Research Institute at the PBC, controllable anonymity was an important feature in the design of the e-yuan. Furthermore, he added that complete anonymity was impractical if the e-yuan was to help prevent criminal behavior such as money laundering, terrorist financing, and tax evasion.[j] The PBC is using what they call a hybrid model, hence a combination of DLT, being used only in limited areas, and other technologies. Nevertheless, their core e-CNY system is a centralized ledger.

➢ From a European perspective the topic is in advanced stages of research and development. In October 2020 the ECB issued for public consultation a Report on Digital Euro[k] (hereinafter "Report" or "Report on a digital euro"). The main reason for undertaking that research is to ensure that consumers continue to have unfettered access to central bank money in a way that meets their needs in the digital age. The main benefits/motivations for issuing a digital Euro pointed out in the report are: (i) to support the digitalization of the European economy; (ii) in response to a significant decline in the role of cash as a means of payment, (iii) to tackle sovereignty concerns related to foreign private digital means of payment in the euro area or possible future foreign CBDC, (iv) could be used as a new monetary policy transmission channel, (v) could be useful to mitigate risks to the normal provision of payment services, (vi) issued to foster the international role of the euro, and (vii) issued to support improvements in the overall costs and ecological footprint of the monetary and payment systems. Two digital euro scenarios are being discussed in the report: 1) account based digital euro, hence by opening accounts directly with the Eurosystem or through supervised intermediaries and 2) bearer digital euro – also referred to as "token-based" or "value-based" digital euro – which would likely require the involvement of supervised intermediaries. Although banknotes and coins are anonymous, this cannot be the case for digital euro since anonymity is not allowed for electronic payments. As expected, in mid-July 2021 the Eurosystem launched the digital euro project, whose investigation phase will last 24 months, although no technical obstacles were identified during the preliminary experimentation phase. From a technological standpoint, the report describes the implications of both options, hence centralized versus decentralized technologies for central banks, intermediary institutions, and end users.[l]

➢ The United Kingdom is also moving forward on this topic. Although the British authorities have not yet decided whether or not to issue a CBDC, the BoE announced on April 2021 the joint creation, together with the HM Treasury, of a CBDC Taskforce to coordinate the exploration of a potential UK CBDC. This announcement comes a year after a Discussion Paper issued by the BoE in March 2020.[m] The Discussion paper pointed out the opportunities that CBDC could present, for the way that the Bank of England achieves its objectives of maintaining monetary and financial stability. These opportunities include supporting a resilient payments landscape, supporting competition, efficiency, and innovation in payments, addressing the consequences of a decline in cash etc. A CBDC Engagement Forum, co-chaired by the BoE and HMT, has also been established. This Forum looks at all aspects of a central bank digital currency apart from the technology it might use and helps the BoE to understand the practical challenges of designing, implementing, and operating a CBDC.[n] From a technological standpoint, although the BoE recognizes the benefits of DLT, it does not exclude the possibility of building its CBDC using more conventional centralized technology. The BoE states that careful attention should be paid to the design features of the CBDC, especially related with the data privacy and protection. Nevertheless, it recognizes that the CBDC should comply with AML rules which rules out full anonymity.

➢ The Federal Reserve System (FED) is in an advanced research phase on CBDC. Since 2020 it has established a multidisciplinary team, with application developers from the Federal Reserve Banks of Cleveland, Dallas, and New York, which supports a policy team at the Federal Reserve Board that is studying the implications of digital currencies on the payments ecosystem, monetary policy, financial

Fig. 7.1 (continued)

stability, banking and finance, and consumer protection.[o] Furthermore, the Federal Reserve Bank of Boston is partnering with Massachusetts Institute of Technology (MIT) to research the feasibility of the core processing of a CBDC, and a white paper has been released in early February 2022. Based on the work so far, the FED's policy considerations on CBDCs converge with those of other jurisdictions and central banks such as improving the payments efficiency; promoting competition and reducing costs; reducing cross-border frictions; preserving financial stability and monetary policy transmission; increasing financial inclusion; protecting privacy and safeguarding financial integrity. On the latter, it recognizes the importance of safeguarding the privacy of households when making payment transactions, but in order to protect and maintain the integrity of the financial system, digital verification of identities should be considered during the design of any CBDC.[p] A recent public discussion paper by the FED specifies that even though no decisions have been made on whether to pursue a CBDC, analysis suggests that a potential U.S. CBDC would best serve the needs of the United States by being privacy-protected, intermediated, widely transferable, and identity-verified. On the latter, the FED acknowledges that the CBDC design features should comply with AML/CFT rules and therefore any intermediary in the process would need to verify the identity of the person accessing CBDC. From a technological standpoint, the FED is exploring several options, including blockchain technology, or the use of DLT for wholesale payments.[q]

[a] Boar and Wehrli 2021.
[b] Project Sand D–llar - A Bahamas Payments System Modernization Initiative, Central Bank of the Bahamas 24 December 2019.
[c] https://www.eccb-centralbank.org/p/about-the-project.
[d] Bergara and Ponce 2018.
[e] https://www.riksbank.se/en-gb/press-and-published/notices-and-press-releases/notices/2021/riksbank-begins-cooperation-with-external-participants-in-e-krona-pilot/.
[f] E-krona pilot Phase 1, Sveriges Riksbank 2021.
[g] https://www.jinse.com/news/blockchain/1003029.html.
[h] https://www.centralbanking.com/central-banks/currency/digital-currencies/7952751/pboc-launches-new-phase-of-e-yuan-pilot.
[i] https://www.bis.org/publ/othp59.htm.
[j] https://thepaypers.com/cryptocurrencies/chinas-central-bank-to-employ-controllable-anonymity-for-the-e-yuan--1248021.
[k] Central Bank 2020.
[l] https://www.ecb.europa.eu/press/pr/date/2021/html/ecb.pr210714~d99198ea23.en.html.
[m] Bank of England 2020.
[n] https://www.bankofengland.co.uk/research/digital-currencies/cbdc-engagement-forum.
[o] Brainard 2020.
[p] Brainard 2021.
[q]: Money and payments The US Dollar, in the age of digital transformation, Federal Reserve System, 2022.

Fig. 7.1 (continued)

banks' balance sheet, CBDC will receive similar treatment. These funds are issued by the central bank either in a physical (cash in circulation) or digital form (CBDC and commercial banks accounts) and as such they fall within the definition of central bank money. On the other hand, the funds that customers have deposited with commercial banks are private money that belongs to them and as such are registered as a liability in the balance sheet of the commercial banks. The creation of central banks liabilities is closely linked with the legal mandate of central banks. But although central banks have been providing a digital form of cash to commercial banks or to the government, they have thus far not provided any to the general public. Therefore, the current "cash in circulation" liability account is related always to the physical cash in circulation rather than to its digital form. As such, relevant central bank laws should clearly authorize,[25] as part of their functions for the creation of central bank liabilities, the issuance of currency in circulation in digital form.

7.6 Legal Tender

Many central banks around the world are the sole legally mandated authority to issue legal tender currency in their jurisdiction. Some of the applicable laws limit the issuance of currency to banknotes and coins only and some others use a broader wording without limiting the issuance of currency to banknotes and coins.[26] By giving legal tender status to the currency they issue, central banks allow the use of the banknotes and coins anywhere in a certain jurisdiction and also as a means of payment where the payee is "obliged" to accept the designated legal tender currency provided by the payer in exchange for the goods or services received.

What about issuing a digital form of legal tender currency? Technically speaking, that could be possible by amending central banks laws and clearly designating CBDC as legal tender currency. But this is not just a matter of legal technicality. A legal tender physical currency is widely accessible and usable by the entire population as a means of payment, which unfortunately could not be the case for a digital form. Hence, the designation as legal tender of a digital currency could be considered as unfair to disadvantaged social groups such as older people, disabled people or people who are digitally or financially excluded,[27] in particular when the adoption of the digital currency is widespread before it brings any type of advantage to its users or is necessary to execute certain transactions. The designation of a legal tender CBDC requires careful assessment of the population cash usage habits beforehand as well as of the consumer protection mechanisms to allow the use of legal tender physical currency in specific situations when that is needed or required. It is important to specify though, that we are speaking about currency in circulation and not about central bank book money which usually does not have legal tender status. The same approach should be followed even with account based CBDC, to avoid the policy implications mentioned above on disadvantaged people. On the other hand, token-based CBDC could be considered a central bank liability with the central bank as the sole issuer. This is particularly important to be legally founded, to avoid repetition of nineteenth century disruptive situations caused by commercial banks issuing banknotes and not honoring their obligations to convert those notes in real currency[28] (a situation that, regrettably, saw a "remake" with the 2022 crypto crash).

---

[25] Such would require an ad hoc analysis, but some existing central bank laws may, due to the applicable definitions of money and currency, allow for the issuance of money in digital form.

[26] Bossu et al. 2020.

[27] For more on this see Riksbank 2019.

[28] Bossu et al. 2020.

7.7 El Salvador Case

To conclude this chapter, and although not entirely related with the topic at hand, we present the country case of El Salvador that, in 2021, adopted Bitcoin as legal tender. This may be useful and instructive since the Salvadorean case presents some aspects that can be relevant for a CBDC implementation: the provision of legal tender directly to the public by the central bank and authorities (such as in the payment of governmental allowances and benefits), the adjustment by the population to "official" digital channels (such as government sponsored wallets), and the granting of legal tender status to a non-physical currency.

7.7.1 Country Context

El Salvador is a small country[29] located in Central America, with a population of 6.5 million inhabitants[30] and the US Dollar as its legal tender currency. The central bank ceased the issuance of the Salvadoran Colon starting January 2001.[31] El Salvador is heavily reliant on remittances, and it leads the region in remittances per capita, with inflows equivalent to nearly all the country's export income. According to World Bank data,[32] the remittances received comprise 24.1% of the GDP in 2020 with an increasing trend in the last five years and the majority of the remittances' funds coming from Salvadorans living in the United States. As such there is a vast USD inflow into El Salvador which is transferred mostly via traditional remittance channels.

The cost of transferring money from the USA to El Salvador has been decreasing over time. According to the World Bank database, in the first quarter of 2021 the average total cost of sending US$ 200.00 was 2.91% of the amount sent compared to 3.13% in the fourth quarter of 2020 and 3.53% in the fourth quarter of 2019.[33] Globally, sending remittances costs an average of 6.38% of the amount sent.[34]

On the other hand, the financial inclusion level in El Salvador is very low. According to the latest Global Findex survey issued in 2021, only 31% of the adult population has an account in a financial institution compared to 29% in 2017, hence a slight increase in financial account ownership figures. This is much lower compared to the global data results, where the share of adults who have an account with a financial institution or through a mobile money service rose from 68 to 74% worldwide.

---

[29] With 8124 mi^2 is roughly the same size as Massachusetts (7838 mi^2) or Slovenia (7827 mi^2).

[30] WorldData.info consulted on 16 November 2022.

[31] https://web.archive.org/web/20070708094105/, http://www.bcr.gob.sv/ingles/integracion/ley.html.

[32] https://data.worldbank.org/indicator/BX.TRF.PWKR.DT.GD.ZS?locations=SV.

[33] https://remittanceprices.worldbank.org/en/corridor/United-States/El-Salvador?start_date=1609477200&end_date=1617249599.

[34] https://remittanceprices.worldbank.org/en.

The main reason cited from the survey respondents (54%) for not having an account was that the financial services were too expensive. Although 70% of the remittance receivers are women, only 24% of them have an account. Most of the remittance household recipients (94%) use the money to cover their daily consumption and around 79% can be classified as poor or at risk of falling into poverty.[35] Therefore, and taking all of the above into consideration, it is clear that Salvadorean population is highly sensitive to costs of financial services to further access and use them.

7.7.2 Bitcoin as Legal Tender

The president of El Salvador announced in June 2021 that he would send proposed legislation to the Congress to make Bitcoin legal tender in the country. Part of the announced reasons behind this decision included the opportunity to promote the formal economy and enhance financial inclusion as well as the possibility to allow Salvadoreans to transfer billions of remittances easier, cheaper, and faster.[36] Inflation pressures to USD seemed to be another reason.

Despite the IMF and World Bank concerns,[37] the new law which was approved by the Congress in June 2021 took effect starting September 2021 and El Salvador became the first jurisdiction in the world to adopt Bitcoin[38] as legal tender. This means that it is now accepted as a means of payment for goods and services throughout the country, unless the payee is unable to provide the technology needed to process the transaction.[39] Incentives were also being applied as the Government offered US$ 30.00 of Bitcoin for every new digital wallet activated.

Nevertheless, one year later, data shows that the use of Bitcoin has not been embraced by the population as it was initially expected by authorities and crypto enthusiasts. The Government claimed in January 2022 that in El Salvador there were 4 million users of Bitcoin[40] but according to the first business survey for 2022 published by the Chamber of Economy and Industry in El Salvador, 91.7% of the respondent answered that the launch of Bitcoin has been indifferent to their business and furthermore, 86.1% related that they have not performed any selling in Bitcoin so far.[41] Hence, even though citizens have opened the government-sponsored bitcoin wallets, their usage has been negligible. According to another study published in

---

[35] https://publications.iadb.org/publications/english/document/Remittance-Recipients-in-El-Salvador-A-Socioeconomic-Profile.pdf.

[36] https://www.npr.org/2021/06/06/1003755600/bitcoin-cryptocurrency-money-el-salvador.

[37] https://www.reuters.com/business/el-salvador-bitcoin-plan-bulletproof-president-says-2021-06-23/

[38] Or any cryptocurrency.

[39] https://www.bbc.com/news/business-57507386.

[40] https://restofworld.org/2022/el-salvador-bitcoin/.

[41] https://camarasal.com/wp-content/uploads/2022/03/SondeoEmpresarial35422_FINAL10032022.pdf.

July 2022 by the National Bureau of Economic Research, only 20% of the firms in El Salvador have accepted bitcoin for payments. Roughly 5% of all sales have been paid in bitcoin through the government sponsored Chivo Wallet, and just as most households using Chivo prefer to keep their money in cash rather than in bitcoin, 88% of firms convert their bitcoin into dollars.[42]

Small business owners and individuals list as main issues the volatility of the virtual currency which made people lose money, logistical problems due to the limited number of available ATMs around the country for cash exchange and technical issues not resolved from the Chivo customer service team.[43] In January 2022, the IMF urged El Salvador to remove Bitcoin as a legal tender.

Despite all the challenges and setbacks, the El Salvador case is not expected to remain isolated. In April 2022, the Central African Republic became the second country in the world to adopt bitcoin as legal tender.[44] In Paraguay, even though the bill regulating crypto mining and trading was passed by the senate in August 2022, it was failed to pass by the Paraguay's lower house in December 2022.[45] Panama has already passed a bill in August 2022 regulating and permitting the use of crypto-assets in the economy as a mean of payment.[46] Outside Latin American countries, Malta is the one to watch. The Maltese have always had reservations concerning the Euro and the country is traditionally welcoming to blockchain and crypto-assets activities and investments.[47]

Nevertheless, considering the skepticism and concerns coming from international bodies, such as the IMF and the World Bank, and the reliance that many countries have on these institutions in terms of financial support and assistance, but also the risks that Bitcoin and other cryptocurrencies may pose due to their volatility and absence of control by the regulators, the decision of implementing Bitcoin or other cryptocurrencies as legal tender must be weighed carefully against the country priorities and the financial wealth of the population. The current status of the crypto-market, in particular following the collapse of Terra Luna[48] and FTX bankruptcy[49] (both followed by other related bankruptcies and a significant downward trend in cryptocurrencies' prices across the board) may act as a deterrent for jurisdictions that were considering adopting Bitcoin as legal tender or moving towards a more crypto-friendly approach. The results of El Salvador's experience, the impacts of

---

[42] https://www.nber.org/digest/202207/el-salvadors-experiment-bitcoin-legal-tender.
[43] https://restofworld.org/2022/el-salvador-bitcoin/.
[44] https://www.bbc.com/news/world-africa-61565485.
[45] https://www.coindesk.com/policy/2022/07/15/paraguayan-senate-passes-bill-regulating-crypto-mining-and-trading/.
[46] https://www.reuters.com/world/americas/panama-lawmakers-pass-bill-regulate-use-crypto-assets-2022-04-28/.
[47] https://www.cityam.com/more-countries-line-up-to-make-bitcoin-legal-tender-which-one-will-be-next/.
[48] https://www.euronews.com/next/2022/05/12/terra-luna-stablecoin-collapse-is-this-the-2008-financial-crash-moment-of-cryptocurrency.
[49] https://www.wsj.com/articles/ftx-collapses-into-bankruptcy-system-that-still-hasnt-figured-out-crypto-11668550688.

FTX implosion on the Bahamian economy,[50] as well as on-going regulatory initiatives such as European Union's MiCA,[51] will no doubt shape this field going forward and have a significant impact in terms of the approaches to be adopted by jurisdictions across the globe.

References

Auer R, Cornelli G, Frost J (2020/ 2021) Rise of the central bank digital currencies: drivers, approaches and technologies. BIS working papers, no. 880
Bank of England (2020) Discussion paper - Central Bank Digital Currency: Opportunities, challenges and design. https://www.bankofengland.co.uk/paper/2020/central-bank-digital-currency-opportunities-challenges-and-design-discussion-paper. Accessed 23 May 2023
Bech M L, Garratt R (2017) Central bank cryptocurrencies. BIS Quarterly Review, September
Bergara M, Ponce J (2018) Central Bank Digital Currency: The Uruguayan e-Peso case. SUERF
Boar C, Wehrli A (2021) Ready, steady, go? – Results of the third BIS survey on central bank digital currency. BIS working papers, no. 114, January
Bossu W, Itatani M, Margulis C, Rossi A, Weenink H, Yoshinaga A (2020) Legal Aspects of Central Bank Digital Currency: Central Bank and Monetary Law Considerations. IMF working papers, November
Kahn C, McAndrews J, Roberds W (2005) Money is Privacy. International Economic Review, Vol. 46, no. 2, Wiley
McAndrews J (2017) The Case for Cash. ADBI Working Paper Series, no. 679, March
Riksbank (2019) The state's role in the payment market. The payments inquiry. Report by the Committee Directive of the Ministry of Finance, Government Offices of Sweden, Stockholm

Other Documents

Brainard L (2020) An update on digital currencies, Federal Reserve Board and Federal Reserve Bank of San Francisco's Innovation Office Hours, San Francisco, California. https://www.federalreserve.gov/newsevents/speech/brainard20200813a.htm. Accessed 23 May 2023
Brainard L (2021) Private Money and Central Bank Money as Payments Go Digital: An Update on CBDCs, Speech, Consensus by CoinDesk 2021 Conference, Washington D.C. https://www.federalreserve.gov/newsevents/speech/brainard20210524a.htm. Accessed 23 May 2023
European Central Bank (2020) Report on Digital Euro. https://www.ecb.europa.eu/paym/digital_euro/report/html/index.en.html. Accessed 23 May 2023

---

[50] FTX had its headquarters and was based in the Bahamas.
[51] Markets in Crypto-assets: https://eur-lex.europa.eu/legal-content/EN/TXT/?uri=CELEX%3A52020PC0593.

Chapter 8
Smart Contracts and the Law

Agata Ferreira

Contents

8.1	Introduction	126
8.2	Origins	127
8.3	Advantages	128
8.4	Legal Debate	131
	8.4.1 Is a Smart Contract a Contract?	131
	8.4.2 Code is Law?	133
8.5	Selected Smart Contracts Legislation	136
	8.5.1 The United States	136
	8.5.2 Europe	138
	8.5.3 United Kingdom	139
8.6	Conclusions	140
References		140

Abstract Smart contracts raise a number of interesting legal issues. An ongoing legal debate concerns a number of potential legal pitfalls related to smart contracts, including the existence of a valid, binding legal contract, the ability to reconcile the immutability of blockchain records with contractual deficiencies, governing law and applicable jurisdiction, consumer protection, anti-money laundering/combating the financing of terrorism (AML/CFT) requirements, implications of automated, unstoppable execution, as well as privacy and confidentiality of smart contracts deployed on borderless decentralized blockchain networks. This chapter provides an overview of the most commonly invoked advantages of smart contracts and highlights the fervor of the scholarly legal debate surrounding smart contracts. It also presents a range of legislative initiatives taken so far in response to the smart contracts phenomenon.

Keywords Smart contracts · Contract law · Automation rights and duties · Private autonomy

---

A. Ferreira (✉)
Department of Administrative Law and Public Policy, Warsaw University of Technology, Faculty of Administration and Social Sciences, Pl. Politechniki 1, Warsaw 00-661, Poland
e-mail: agata.ferreira@pw.edu.pl

8.1 Introduction

Smart contracts are being deployed across a range of sectors, including the financial sector, public sector, supply chain management, and the automobile, real estate, insurance, and health care industries.[1] Global smart contracts' market size has been rapidly growing and is expected to gain a compound annual market growth rate of 17.4% in the forecast period of 2020 to 2025, and is expected to reach USD 208.3 million by 2025.[2] Smart contracts have been vigorously debated by legal scholars and motivated mixed views. Some scholars call smart contracts 'one of the first truly disruptive technological advancements to the practice of law since the invention of the printing press',[3] others call them a 'dumb idea'.[4] Legislators and regulators have also begun paying attention and the first legislative initiatives addressing smart contracts started to emerge.

Smart contracts raise a number of interesting legal issues despite contract laws being a fairly well-developed area of law and contractual freedom being widely recognized across most jurisdictions.[5] It is difficult and 'probably impossible'[6] to define smart contracts and there is no 'universally accepted'[7] definition. Equally, there is no unified, structured, and systematic classification of smart contracts.[8] An ongoing legal debate concerns a number of potential legal pitfalls related to smart contracts, including the existence of a valid, binding legal contract,[9] the ability to reconcile the immutability of blockchain records with contractual deficiencies, governing law and applicable jurisdiction, consumer protection, AML/CFT requirements, implications of automated, unstoppable execution, as well as privacy and confidentiality of smart contracts deployed on borderless decentralized blockchain networks.[10] This chapter provides an overview of the most commonly invoked advantages of smart contracts and highlights the fervor of the scholarly legal debate surrounding smart contracts.

---

[1] Chamber of Digital Commerce 2016.
[2] Marketsandresearch.biz 2020.
[3] Wright and De Filippi 2015, p 10.
[4] O´Hara 2017.
[5] See Verstraete 2019 for a critique of the optimism about smart contracts; Mik 2019 for a discussion on the limitation of smart contracts and their limited potential impact on contract law and the legal system; de Caria 2019 on the general legal dimension of legal contracts; Madir 2018 for a discussion about issues relating to enforceability of smart contracts, authentication of signatures, liability allocation, governing law, fraud risks and data protection; Low and Mik 2019 for a critical evaluation of the popular claims surrounding the potential of blockchain technologies to disrupt the legal system; Jaccard 2017 for an assessment of the legal relevance of smart contracts; Meyer 2020 for a legal analysis of the unwinding and termination of smart contracts.
[6] Rühl 2020, p 2.
[7] Meyer 2020.
[8] For the foundation for smart contracts' classification, see Tönnissen and Teuteberg 2018; for the classification of strong and weak smart contracts, see Raskin 2017.
[9] Levi and Lipton 2018; Raskin 2017.
[10] De Caria 2019; Meyer 2020; Rühl 2020.

It also presents a range of legislative initiatives taken so far in response to the smart contracts phenomenon.

The first section introduces the origins of smart contracts and their revival with the dawn of blockchain technology. The second section explains most commonly invoked benefits of smart contracts and examines their merits. The third section provides an overview of the ongoing legal scholarship debate. The fourth section analyses selected smart contracts legislative initiatives across various jurisdictions. The final section offers concluding remarks.

8.2 Origins

Nick Szabo, a computer scientist and legal theorist, first coined the phrase "smart contracts" in the 1990s and gave an example of a vending machine as a 'canonical real-life example, which we might consider to be the primitive ancestor of smart contracts'.[11] A vending machine operates on the basis of a pre-programmed *if–then* logic, where payment triggers irrevocable automatic action, which cannot be stopped once the coin has been inserted. Szabo proposed one of the first definitions of a smart contract: 'a set of promises, specified in digital form, including protocols within which the parties perform on these promises'.[12]

Smart contracts pre-exist blockchain technology, but blockchain has enabled the progress of smart contracts from simple automated contracts to fully autonomous self-executing and self- performing contracts built on decentralized platforms and supported by a blockchain ecosystem.[13] Blockchain enhances smart contracts with its unique characteristics that guarantee their integrity and security. Blockchain, as it is known today, has been brought to life by the 2008 Bitcoin manifesto by the mysterious Satoshi Nakamoto, whose identity remains unknown.[14] Explained simply, blockchain technology is a specific type of distributed database, where data is stored in an immutable chain of blocks chronologically linked with each other and secured against tampering by cryptography. Every time a new block is added to the chain, it is validated by nodes in accordance with pre-programmed consensus protocol. Conceptually, blockchain provides one set of agreed on, reliable, and verified records of data and eliminates the need for any third party, intermediary, or other authority to verify or store the records. Bitcoin network is the first application of blockchain technology as theorized by Nakamoto and cryptocurrencies constitute the most common use of blockchain. However, platforms such as Ethereum, that support Turing complete programming, enabled wide use of blockchain technology and significantly increased its potential beyond cryptocurrencies, including building smart contracts (although building smart contracts on a Bitcoin network has also

---

[11] Szabo 1996.
[12] Ibid.
[13] Lauslahti et al. 2017.
[14] Nakamoto 2008.

been possible to a limited extent). It became possible to implement *if–then* business logic on a blockchain to enable autonomous self-executing and self-performing computerized scripts, called, rather confusingly, smart contracts. As soon as they gained popularity, they caught the attention of the legal community, which began to question just how 'smart' they are and whether they are at all 'contracts'.

8.3 Advantages

Smart contracts' enthusiasts emphasize the benefits of efficiency resulting from the attributes of automation, standardization, and transparency.[15] Blockchain-based smart contracts enable, for example, automated insurance claims, transparent and efficient supply chains, and corporate governance. Costs and time savings and reduction of operational and counterparty risks resulting from the use of smart contracts enabling peer-to-peer transactions also contribute to efficiency gains.[16] The opposite view is that the rigid and inflexible code impoverishes contractual relationships and in fact reduces contractual efficiency as it is too delimited and finite to reflect the flexibility and efficiency of traditional contractual relationships 'driven by the richness of semantic expression and the power of human judgment'.[17] Programming contractual terms representing implied principles of contract law or ambiguous legal terms like, 'reasonable efforts', 'best endeavors', or 'good faith' into *if–then* computerized logic is a challenge. Capturing by code all *ex-post* circumstances provided by law that allow contractual party to excuse themselves from contract performance without incurring liability, like *force majeure*, illegality or public policy (depending on the legal system), may be incompatible with *ex-ante* regulation of contractual relationships by smart contracts.[18] On the other hand, an overly complicated code encapsulating complex contractual terms could be prone to errors, and the costs of technical maintenance, testing, and fixing bugs could outweigh any efficiency gains.[19] Smart contract platforms could make forming certain contractual relationships more difficult, more expensive, and less efficient as it requires a certain degree of technical sophistication and technical resources and support. Smart contracts are more likely to result in improved efficiency in the industries with high automation and high-volume transactions flows, in sectors relying on standardized contractual terms and repetitive transactions[20] and where *if–then* logic can be easily and objectively ascertained, for example, in the derivatives market.[21] For contractual environments where there

---

[15] Di Angelo et al. 2019, p 392.
[16] Madir 2018, p 1.
[17] Sklaroff 2018, p 286.
[18] Tjong Tjin Tai 2018.
[19] Verstraete 2019, p 789.
[20] Madir 2018, p 4; Cieplak and Leefatt 2017.
[21] International Swaps and Derivatives Association 2019.

is *ex-post* uncertainty, or where parties prefer to avoid drafting highly customized agreements, the use of smart contracts may not result in efficiency gains.[22]

Smart contracts' optimists envision commercial activity through the exclusive use of smart contracts 'avoiding the high costs of contract drafting, judicial intervention, opportunistic behavior, and the inherent ambiguities of written language'.[23] Instead of relying on time-consuming and costly judicial enforcement, automatic performance of smart contracts potentially eliminates the need for institutional enforcement and presents a cheaper and more effective alternative. Traditional *ex-post* judicial enforcement could be replaced by *ex-ante* guarantee of performance by automated self-executing smart contracts. Some argue that this does eliminate disputes between the parties as to the correct performance of contractual terms or their breach since the blockchain does not ensure that the code correctly represents the terms of the agreement between the parties.[24] Guaranteeing contract performance by automatically executing code could create opportunities for contractual relationships between parties that would have difficulties trusting each other in traditional contracting. Shifting trust onto technology facilitates transactions not only between the parties that do not know each other, but also between the parties that know each other but would be reluctant to transact with each other in the absence of technological means of guaranteeing performance. The use of non-ambiguous computer code could also potentially eliminate any disputes resulting from linguistic ambiguities of traditional contracts, on the assumption that 'written in programming language, smart contract remove the ambiguity inherent to natural language'.[25] Grimmelmann questions unambiguity argument and claims that 'smart contracts do not eliminate ambiguity—they hide it' as 'perfect unambiguity is impossible even in theory, because the technical layer ultimately rests on a social one'.[26] The social institutions, which determine smart contracts semantics and on which smart contracts depend are 'those that establish and limit blockchain communities'. Ultimately, 'blockchain is made out of people' and so is smart contracts programming language and underlying blockchain consensus. The 'DAO hack' is a perfect example of split semantics and a dilemma: 'What matters more: What the code said or what people thought it said?'.[27] The DAO, distributed autonomous organization, was an Ethereum blockchain-based online venture-capital fund, designed to be managed through smart contracts. After an unidentified Ethereum user was able to exploit the code and transfer $60 million worth of Ether out of DAO, the DAO community debated whether it was a hack, or a legitimate action allowed by the code. As a result, a large majority of Ethereum users opted for reversing the transfers and returning the funds to the investors. It led to a hard fork in the Ethereum network and a split into two different blockchains, with two different and incompatible semantics, Ethereum and Ethereum Classic. Despite

---

[22] Sklaroff 2018, p 291.
[23] Sklaroff 2018, p 263.
[24] Mik 2019.
[25] Finck 2019, p 27.
[26] Grimmelmann 2019.
[27] Levine 2016.

their potential to reduce costs, smart contracts can be expensive to implement in the first place, because of 'ex ante information costs to determine all contingencies'[28] and the costs of switching to smart contracts network and persuading counterparties to participate in that network. Automation and minimization of human intervention could also result in collateral costs related to any coding errors, implications of immutability, and any potential need to reverse unintended transactions.

For libertarians, smart contracts are considered an embodiment of individual self-determination and autonomy due to their self-contained rules and the ability to contract free from state interference.[29] The state cannot stop a smart contract automated program, influence a running smart contract script in decentralized system, or shut down an autonomous software agent.[30] On a conceptual level, blockchain technology enables personal autonomy free from 'inefficient and corruptible institutions' and 'an insistence on the primacy and desirability of private social ordering, and frustration with the law and lawyers'.[31] Verstraete opposes this view and calls illusory the perception of smart contracts as 'a political system that deduces outcomes simply in reference to the contract' and therefore as more adequately preserving the autonomy of the parties.[32] The code might be the greatest promise for libertarians, but even Lessig warns that it is also the greatest threat because the code is deterministic, binary matrix of yes or no, with no middle ground, and might be built to protect certain values, or to do the opposite, dismiss them.[33]

Fairfield notes disintermediation of smart contracts shifts consumer protection from courts, which have a history of 'longstanding refusal to enforce contract terms proffered by consumer', to automated consumer-grade purchasing agents.[34] He calls disintermediation a 'radical disenfranchisement of consumers in online contracting environments'.[35] 'Smart contracts are designed to test the poorly conceived legal foundations of the current mass-market consumer- term-exclusion regime'.[36] The opposite view is that smart contracts might not be appropriate for consumer contracts unless there are measures implemented that allow consumers to understand coded smart contracts and their implications, for example, translation of smart contracts into a readable consumer friendly format. In many jurisdictions, consumers are granted informational rights, cooling-off periods, and rights to revoke the contract within a specific time frame. Automation might also reverse the burden of proof to the consumer's disadvantage. Retailers would receive the payment through automatic smart contract execution, and consumers would have the burden to prove if they consider that such payment should not have been made. The shifting of such risk onto

---

[28] Werbach and Cornell 2017, p 360.
[29] Kaal and Calcaterra 2017; Raskin 2017, p 335.
[30] De Filippi and Hassan 2016.
[31] Sklaroff 2018, p 268.
[32] Verstraete 2019, p 789.
[33] Lessig 2006, p 6.
[34] Fairfield 2014, p 39.
[35] Ibid, p 41.
[36] Ibid.

the consumer might not be appropriate or acceptable in the current era of laws and regulations being introduced to protect consumers and individuals against increasingly complex digital goods and services.[37] For example, Illinois implemented new blockchain laws that disallow blockchain records for consumer credit defaults, utility cut-offs, health insurance coverage changes, or a recall of a product.[38] The European Parliament Committee of Legal Affairs also highlights in its report a concern over the weaker party to smart contracts and the need to ensure mechanisms that could stop the execution of smart contracts to protect vulnerable party.[39]

8.4 Legal Debate

The idea of a self-executing software that automatically and autonomously implements contractual terms on a peer-to-peer and immutable basis remains legally problematic. Legal debates are often prompted by claims that code should be treated as a 'foreign legal system',[40] that the technology will subject the provision of justice to market forces and break the state's monopoly over the court system'[41] and that 'smart contracts don't have a need in a legal system to exist, [and] they may operate without any overreaching legal framework' as 'they represent a technological alternative to the whole legal system'.[42] Smart contracts are thought to be providing new contractual forms alongside traditional ones or even replacing contract law altogether.[43] Smart contracts' optimists are quickly countered by those who point to the limitations and deficiencies and call the optimistic version for smart contracts' role 'illusory'.[44]

8.4.1 Is a Smart Contract a Contract?

Legal community has been grappling with the question of whether smart contracts are contracts in a legal sense. Much of this confusion is due to terminological misconceptions and an overlap of law and technology, which resulted in intense scholarly debates. Some of them produced a rebuttal of smart contracts as neither smart nor contracts.[45] The views and opinions on whether smart contracts can be classified

---

[37] Clifford Chance and European Bank for Reconstruction and Development 2018, p 5.
[38] Section 15(d) of Illinois Public Act 101–0514.
[39] European Parliament 2020.
[40] Reyes 2017, p 389.
[41] Raskin 2017, p 335.
[42] Savelyev 2016, p 21.
[43] De Filippi and Hassan 2016.
[44] Verstraete 2019, p 774.
[45] Felten 2017.

as legal contracts differ. Kolber explains that a code can stipulate the terms of the contract, just as paper does in the case of a written contract. However, in his view, 'neither code nor paper are contracts because contracts are not physical entities' therefore the statement that code is the contract is simply false.[46] The sceptics maintain that smart contracts do not amount to legal contracts per se and that 'without a form of legal existence or recognition, a smart contract is simply a non-binding piece of computer code that claims the proof of something' and can be recognized as legally relevant only to the degree that they create 'normative means to regulate social behavior, which competes with the scope of application of legal regulation'.[47] Meyer compares smart contracts to a 'technical aid, an "app" so to speak'; and therefore not contracts in the legal sense but a 'computer babble'—computer jargon based on legal language that has no legal implications and is therefore not a contract.[48] Some theorists, like Savelyev, go further by saying that a 'smart contract does not give rise to a legal bond between the parties' and instead there is a 'technical bond' and 'such a bond is much more solid than a legal one'.[49] Werbach and Cornell are also skeptical about the legal implications of smart contracts and claim that 'just because smart contracts are being implemented today on the exotic technology of the blockchain does not mean they raise novel or interesting legal issues' as they are just 'technological manifestations of familiar contractual processes: escrow and self-help'.[50] There is no common understanding of what exactly the relation between smart contracts and traditional legal contracts is.

Some authors believe, that unlike traditional contracts, smart contracts follow simple conditional *if–then* logic, whereas the matrix of traditional contracts is based on consensual connection, a meeting of the minds, within a binary sequence of the 'offer and acceptance'.[51] According to Meyer, this difference raises a question of the relationship of smart contracts with traditional concepts of contract law. For others, the conditional framework is common for smart contracts and contract law, where 'promises are made in exchange of other promises'.[52] The use of the word 'contract' often confuses legal community because it has specific implications in law.[53] Szabo coined the term 'smart contracts' and described that objectives and principles for the design of smart contracts systems are derived from 'legal principles', implying solid legal foundations.[54] Although it is now largely accepted that not all smart contracts have legal relevance, technical community often opposes to the use of this terminology because the technical concept has been characterized by the use of legal

---

[46] Kolber 2019.
[47] Jaccard 2017, p 8.
[48] Meyer 2020.
[49] Savelyev 2016.
[50] Werbach and Cornell 2017, p 344.
[51] Meyer 2020.
[52] Catchlove 2017.
[53] Marino 2016.
[54] Szabo 1997.

terms[55] and 'it muddies the waters and invites unnecessary legal/regulatory scrutiny' and causes '(1) chaos, (2) incoherence, (3) unnecessary complexity, (4) regulatory blowback, (5) added costs, (6) confusion, and (7) reduced innovation'.[56] Even Vitalik Buterin, one of the founders of the Ethereum network, said he regretted adopting the term 'smart contracts' and said he should have called them something more boring and technical, perhaps something like 'persistent scripts'.[57]

Contract law fundamentals of each legal system determine what is required to create a valid and binding contract. Common law, like English law, requires three elements: offer, acceptance, and consideration for valid contract formation. The UK Jurisdiction Task Force (UKJT) has recently stated that a 'smart contract is capable of satisfying those requirements just as well as a more traditional or natural language contract, and a smart contract is therefore capable of having contractual force'.[58] In a civil law jurisdiction, using German law as an example, a contract is created when parties reach an agreement on the essential terms by two corresponding declarations of intent (offer and acceptance). General principles of contract law can be applied to determine whether a smart contract is legally valid, binding, and enforceable. Among different types and models of smart contracts, depending on their legal relevance, context, or technical properties, there are smart contracts which have no legal implications and do not amount to legally binding contracts. Non-legal smart contract would be a piece of software that follows *if–then* logic and executes pre-programmed tasks that do not involve triggering a legal relationship. In other words, 'smart contracts' are not contracts in the legal sense, although nothing prevents them from having legal effects.[59]

8.4.2 Code is Law?

'Code is law' became a libertarian ideal upheld by some smart contracts' proponents, according to which technology alone provides sufficient regulatory parameters to govern smart contracts. The phrase 'code is law' was coined by Lessig and has since been used to denote the notion that the underlying software code becomes de facto a regulator of the architecture and infrastructure of cyberspace. Code as a regulator of cyberspace is not a new idea. Self-regulation of cyberspace is described by Reidenberg[60] as Lex Informatica, a concept based on internet self-regulation by the use of relevant 'architecture standards' for a distinct jurisdiction of a network, where the source of governing rules is technology and a network of communities of users,

---

[55] Mik 2017.
[56] CleanApp 2018.
[57] https://twitter.com/VitalikButerin/status/1051160932699770882?s=20. Accessed 25 September 2020.
[58] UK Jurisdiction Taskforce 2019, p 8.
[59] Mik 2019.
[60] Reidenberg 1998.

developers, and other participants, and not legislators and law, and where enforcement is performed automatically, not by courts. One of the main motivations for the creation of an Ethereum network has been to facilitate transactions between individuals with no means of trusting each other because of the 'incompatibility, incompetence, unwillingness, expense, uncertainty, inconvenience, or corruption of existing legal systems'.[61] In this sense, 'Ethereum may be seen as a general implementation of such a crypto-law system', providing an improved, 'upgraded' governance matrix alternative to established legal systems, which became deficient.[62]

Mik counters the proposition that 'code is law' as applicable to smart contracts by saying that such ideas 'falsely assume that it is technically feasible to encode every aspect of a transaction and that it is legally permissible to exclude the operation of the legal system'.[63] In Rühl's view, smart contracts 'need a legal system as a normative point of reference' because it is the law that determines whether there is an enforceable obligation. The code in itself is not law and should not be, and a smart contract is a mere piece of code that does only what is has been programmed to do.[64] Even Lessig, whilst acknowledging that 'we live in an era fundamentally skeptical about self-government', admitted that 'this is an indulgence that is dangerous at any time' and 'particularly dangerous now'. He seems to acknowledge is that the code is a mere organizer of the cyberspace, provider of a set of rules, definer of the ordinance, and an operating language so the speak rather than a legal authority.[65]

Following the emergence of blockchain technology, De Filippi and Wright took the notion 'code is law' one step further, by introducing an idea of Lex Cryptographia, a new paradigm of law representing the 'world where ideals of individual freedom and emancipation might come true'.[66]

Blockchain based smart contracts programmed by using Turing complete programming script, which allows any natural language expression to be translated into a code, brought a new dimension to the notion of cyberspace self-regulation. Lex Cryptographia is being defined as a new subset of law, as 'rules administered through self-executing smart contracts and decentralized (autonomous) organizations. In this model, 'law and code may merge, so that the only way for people to infringe the law is to effectively break the code'. To facilitate assimilation of law into code, laws could evolve towards code-like expressions, which could lead to 'law progressively turning into code'.[67] Some even hail blockchain-based smart contracts as the beginning of the end of classic contract law.[68]

---

[61] Wood 2014, p 1.
[62] Ibid, p 2.
[63] Mik 2019, p 21.
[64] Rühl 2020, p 4.
[65] Lessig 2000.
[66] Wright and De Filippi 2015, p 56.
[67] De Filippi and Hassan 2016.
[68] Savelyev 2016.

In response to that proposition, other scholars find it necessary to 'remind the basic principles of contract law' and to acknowledge smart contracts' 'limited potential impact on contract or the legal system in general'.[69] In fact, even Lex Cryptographia proponents, Wright and De Filippi, admit that mainstream blockchain applications are unlikely to eliminate law and governments, but might 'shift the balance between law and architecture, requiring alternative regulatory mechanisms to successfully manage society'.[70] Blockchain is a transformative technology, capable of re-configuring established economic, legal, institutional, monetary and broader socio-political relationships.[71] Blockchain-based smart contracts purport to represent a new form of social organization. Normative legitimacy of regulatory properties of smart contracts is being questioned, even though, or perhaps because of the fact that, 'code and law may generate the same regulatory effects.[72] The concern is that the governing authority of smart contracts is derived from technology, not law. Smart contracts are perceived as a form of coercion due to their distinct blockchain-based characteristics of automated performance and immutability of records. They do not provide adequate procedures for selection, modification, and validation of background rules of contract law, the nature of which is undetermined. As a result, the rules governing smart contracts are created by omission and fail to account for and distinguish between rudimentary doctrines of contract law.[73] The distribution of wealth and power resulting from such 'overridden' contract law fundamentals and the distribution of interests and obligations between smart contracts platforms participants in a blockchain network governance structure are considered normatively suspect and since smart contracts relegate legal regulation, they warrant the presumption of 'normative deficiency unless proven otherwise'.[74]

Smart contracts also highlight the tensions and have the potential to shift the boundaries between public and private domain. They enable novel forms of economic exchange between individuals, which may challenge the established transacting forms and test the established divide between private and public domain. Karlstrøm states that this unfettered freedom to transact when subjected to blockchain code may have in fact an opposite, disempowering effect of commercial relations becoming 'rigid, irreversible and non-negotiable'. Technology like blockchain may render social relations 'increasingly rigid, at the cost of a loss of dynamism and consequently of a sense of freedom and responsibility'.[75] Blockchain based smart contract systems also lack any conception of common good, common interest beyond its functionality of enabling peer to peer transactions according to the protocol rules. The underlying philosophy of the blockchain model of governance contradicts established social and institutional design, by being 'strongly aligned to anarchist and libertarian

---

[69] Mik 2019, p 22.
[70] Wright and De Filippi 2015, p 51.
[71] Reijers et al. 2016; Reijers and Coeckelbergh 2018.
[72] Verstraete 2019, p 794.
[73] Ibid.
[74] Verstraete 2019, p 795.
[75] Karlstrøm 2014.

theories of social order' and a classical formulation of libertarianism denies any possibility for the state to interfere in the freedom of individuals.[76]

8.5 Selected Smart Contracts Legislation

8.5.1 The United States

Despite an increasing interest in and recognition of smart contracts, only a handful of countries have formulated a regulatory response to smart contracts. In the US, smart contracts legislative initiatives by US states are relatively narrow and govern only selected issues, mostly limited to defining smart contracts and recognition of their electronic form and signatures. Until April 2019, various US states have considered 133 blockchain technology related laws. However, only 7 related to smart contracts and only a few of those have been passed successfully.[77]

Arizona was the pioneer of smart contracts legislation in the United States.[78] In 2017, Arizona passed a law (Bill HB 2417)[79] amending the Arizona Electronic Transactions Act that clarified the use of blockchain technology, including smart contracts, in commercial transactions. The new law has a declaratory character, provides definitions of blockchain and smart contracts, and recognizes them within the existing legal framework as another electronic form of commercial transacting, acknowledging their legal status, validity, and enforceability. The act defines a smart contract as an 'event driven program, with state, that runs on a distributed, decentralized, shared and replicated ledger that can take custody over and instruct transfer of assets on that ledger'. Tennessee followed suit and passed similar legislation in March 2018 (Bill SB 1662)[80] to provide legal recognition to smart contracts in the marketplace. Tennessee's act acknowledges that smart contracts are valid and enforceable under state law, 'may exist in commerce' and cannot be denied legal effect, validity, or enforceability. Both the Arizona's and Tennessee's blockchain laws aim to incorporate smart contracts into existing legal frameworks governing electronic signatures and contract law. Also, the Arkansas legislation of 2019 (Bill HB 1944)[81] and the North Dakota new bill (Bill 1045) largely follows Arizona's and Tennessee's laws, and several other states are considering similar bills to facilitate the development of blockchain technology and recognition of smart contracts.

---

[76] Reijers et al. 2016, 139.
[77] Arcari 2019, p 366.
[78] Neuburger 2017.
[79] https://www.azleg.gov/legtext/53leg/1r/bills/hb2417p.pdf. Accessed 25 September 2020.
[80] https://legiscan.com/TN/text/SB1662/id/1691046. Accessed 25 September 2020.
[81] https://trackbill.com/bill/arkansas-house-bill-1944-hb1944-concerning-blockchain-technology/1733271/. Accessed 25 September 2020.

Few states stand out as more progressive and address smart contracts in a more comprehensive way. For example, Nevada enacted a bill (Bill 398)[82] in June 2017, which amends the Uniform Electronic Transactions Act and allows smart contracts, records, or signatures created, stored, or verified on a blockchain to be enforceable and admissible in evidence. Nevada is also the first US state to ban local governments from imposing taxes on blockchain use. Nevada's bill closely resembles the recent Illinois' Blockchain Technology Act (BTA)[83] implemented on 1 January 2020. Unlike the fragmented approach of other US states, the BTA addresses all three issues of smart contracts: enforceability, admissibility to evidence, and local regulations. Smart contracts are defined simply as contracts stored as electronic records, which are verified by the use of a blockchain. Smart contracts, records, or signatures may not be denied legal effect or enforceability nor be excluded in proceedings as evidence solely because a blockchain was used to create, store, or verify them (similar law regarding the evidentiary effect of smart contracts has been passed in Vermont). The BTA also includes limitations on the use of blockchain records and smart contracts. In certain public policy circumstances, like cancelling a public utility, a notice of default or eviction, a cancellation of health benefits, or a recall of products, notice or confirmation of receipt of notice may not be sent over blockchain.

Finally, BTA prohibits local governments from imposing any limits or requirements on the use of blockchain or smart contracts, including taxes, fees, or certification requirements. The BTA provides a high degree of legal clarity to the use of smart contracts and clarifies their legal status and effect. Finally, Wyoming stands out with 13 blockchain laws enacted to provide a comprehensive legal framework designed to foster innovation and promote the blockchain industry. It has many novel features and defines a smart contract as an automated transaction or any substantially similar analogue, which is comprised of code, script or a programming language that executes the terms of an agreement, and which may include taking custody of and transferring an asset, or issuing executable instructions for these actions, based on the occurrence or non-occurrence of specified conditions.[84]

Smart contract legislative initiatives in US states have been criticized for their hasty passage and unforeseeable effects.[85] Some critics have claimed that these initiatives amount to no more than a promotion of a particular jurisdiction and create a risk of regulatory fragmentation, piecemeal smart contracts legislation, potentially prejudicing harmonizing federal legislation in the future. They are thought to 'demonstrate a premature accommodation of smart contracts and an incomplete attempt to predict the future of technological development'.[86] Given the cross-border character

---

[82] https://www.leg.state.nv.us/Session/79th2017/Bills/SB/SB398_EN.pdf. Accessed 25 September 2020.

[83] http://www.ilga.gov/legislation/ilcs/ilcs3.asp?ActID=4030&ChapterID=20. Accessed 25 September 2020.

[84] Long 2019.

[85] Grenon 2019.

[86] Ibid, p 2.

of blockchain technology, legislative divergence would be confusing and would add compliance costs.[87]

Introduction of multiple definitions of smart contracts and expressly giving them the legal status can also contribute to uncertainties around established legal doctrines and application of existing federal regulations, such as the Uniform Electronic Transaction Act, which is often considered as already sufficient for electronic signatures and smart contracts.[88] The US Chamber of Digital Commerce expressed the view that the existing US law 'supports the formation and enforceability of smart contracts under state law' and that 'additional state legislation, inconsistently drafted, will confuse the marketplace and potentially hinder innovation'.[89] Under US law, largely derived from the common law system, the judiciary will apply the existing common law and statutory contract law sources to determine enforceability of smart contracts even without express legislative recognition of smart contracts. However, with the statutory recognition, courts and others will be less likely to reject smart contracts' enforceability.

Two US federal regulatory and supervisory agencies, the US Commodity Futures Trading Commission (CFTC) and the US Securities and Exchange Commission, also address smart contracts through their investigations and clarifying guidance. The CFTC even issued a primer on smart contracts addressing the benefits, risks, and legal issues of smart contracts.[90] The CFTC claims that a smart contract could be a binding legal contract, depending on the facts and circumstances and explains that existing legal frameworks apply and that smart contracts may be subject to a variety of legislation, depending on their application or product characterization. The primer also emphasizes that, despite many benefits of smart contracts, there are also several risks and legal considerations that must be addressed, including operational, technical, fraud and cybersecurity risks and risks arising out of governance protocols.

8.5.2 Europe

The first country in the world to pass regulations on smart contracts is Belarus. A presidential decree from December 2017 defines a smart contract as a program code intended to function in the transaction block ledger (blockchain), or another distributed information system, for the purposes of automated performance and execution of transactions or performance of other legally significant actions. The decree granted the right to conduct performance and execution of transactions by means of smart contracts in the special zone called the Park of High Technologies. A Russian legislator introduced digital assets into the Russian civil law system by amending the Russian Civil Code in 2019 to include the concepts of digital right and

---

[87] Kim and Boring 2018; Orcutt 2018.
[88] Ibid.
[89] https://digitalchamber.org/policy-positions/smart-contracts/. Accessed 25 September 2020.
[90] US Commodity Futures Trading Commission 2018.

smart contracts. Smart contracts are recognized as means of a contract performance, not as a special type of contract, hence they can be now legally applied to any type of transaction in Russia.[91]

Italy is the first European country to regulate distributed ledger technology and smart contracts in a general manner without reference to any specific industry or application.[92] The new law (No. 12/2019) introduces qualified recognition of smart contract and defines a smart contract as 'a computer program that works through distributed ledger technology and whose performance automatically binds two or more parties based on effects defined by the parties themselves'. Before a smart contract is recognized as an automatable and enforceable agreement concluded in writing, an advanced electronic signature or a qualified electronic signature system is required that enables the identification of the parties.

At the European Union level, an uncertainty concerning the legality of smart contracts and their enforceability in cross-border situations has also been recognized and the draft report of the Committee of Legal Affairs emphasized that the upcoming Digital Services Act provides an opportunity to assess the requirements for smart contracts to be considered legally valid. The Committee called on the European Commission to assess the legality and enforcement of smart contracts in cross-border situations, and to make proposals for the appropriate legal framework. The report strongly recommends that smart contracts include mechanisms that can halt their execution to protect weaker parties and ensure that the rights of creditors in insolvency and restructuring are respected.[93]

8.5.3 United Kingdom

The UKJT published a statement in December 2019 on the status of crypto-assets and smart contracts under the law of England and Wales.[94] The statement concludes that smart contracts are capable of forming valid, binding, and enforceable contracts. It emphasizes the adaptability and flexibility of common law that is capable of catering to technological advancements such as smart contracts. The UKJT explains that there is no reason that the three requirements for contract formation under the common law—offer, acceptance, and consideration- could not be satisfied by a smart contract. Smart contracts can be implemented and interpreted under the existing established legal frameworks and contract law doctrines. Although the legal statement is formed on the basis of English law, other common law legal systems share similarities, particularly in terms of their flexibility and the principles of contract formation.

---

[91] Moyle and Bacheyeva 2019.
[92] Rumi et al. 2020.
[93] European Parliament 2020.
[94] UK Jurisdiction Taskforce 2019.

8.6 Conclusions

Smart contracts generate significant interest, have substantial potential, and offer major benefits. The potential of smart contracts has been recognized by the industry, legal community and recently by some regulators and legislators as well. UKJT confirmed that 'in legal terms, crypto-assets and smart contracts undoubtedly represent the future',[95] even though some legal scholars consider them a 'damp, and regrettably widely distributed, squib' of a promised blockchain revolution.[96] It is therefore crucial to understand and clarify the normative dimension of this innovation and develop its legal underpinnings.

Smart contracts are not unified. They vary from simple, straightforward, standardized payment instructions to sophisticated instruments capable of the autonomous performance of a complicated sequence of actions. With the increased use of smart contracts more legal clarification, legislative recognition and regulatory guidance will be needed. The intersection of technology and law presents challenges and, although regulating smart contracts is 'an exceptionally difficult task',[97] adequate regulatory response is essential to provide market confidence and legal certainty to the industry, investors, and consumers.

Several aspects of smart contracts challenge the legal community, but they are unlikely to undermine established legal systems and in principle there seem to be no major obstacles for smart contracts use and no dramatic legal reforms necessary to embrace them under most legal systems. The legislators' response has been limited so far and tends to address basic definitions, legal recognition, and evidential admissibility of smart contracts. Smart contracts should remain firmly on the agenda of the legislators, who should thoroughly consider this technological development and monitor it closely, stepping in to provide legal clarity, support innovation and mitigate risks.

References

Akilo D (2020) Illinois Blockchain Bill to Legalize Smart Contracts and Promote Blockchain Adoption. Business blockchain HQ. https://businessblockchainhq.com/business-blockchain-news/illinois-blockchain-bill-to-legalize-smart-contracts-and-promote- blockchain-adoption/. Accessed 25 September 2020

Angelo MD, Soare A, Salzer G (2019) Smart contracts in view of the civil code. Proceedings of the 34th ACM/SIGAPP Symposium on Applied Computing. https://publik.tuwien.ac.at/files/publik_278278.pdf. Accessed 25 September 2020

Arcari J (2019) Decoding Smart Contracts: Technology, Legitimacy, & Legislative Uniformity. Fordham Journal of Corporate & Financial Law 24(92): 363

---

[95] UK Jurisdiction Taskforce 2019, p 3.
[96] Mik 2019, p 32.
[97] Woebbeking 2019, p 113.

Boto K (2019) The UK Provides Legal Certainty For Smart Contracts And Crypto-assets In Its Landmark Legal Statement. Montaq. https://www.mondaq.com/uk/fin-tech/868000/the-uk-provides-legal-certainty-for-smart-contracts-and-cryptoassets-in-its-landmark-legal-statement. Accessed 25 September 2020

Brett J (2020) Congress Has Now Introduced 32 Crypto and Blockchain Bills. Forbes. https://www.forbes.com/sites/jasonbrett/2020/04/28/congress-has-introduced-32-crypto-%20and-blockchain-bills-for-consideration-in-2019-2020/?sh=262d7c3f786f. Accessed 25 September 2020

Catchlove P (2017) Smart Contracts: A New Era of Contract Use. https://doi.org/10.2139/ssrn.3090226

Chamber of Digital Commerce (2016) Smart Contracts: 12 Use Cases for Business & Beyond a Technology

Cieplak J, Leefatt S (2017) Smart Contracts: A Smart Way To Automate Performance. Georgetown Law Technology Review 1:417-427

CleanApp (2018) Why's Szabo Afraid of "Smart Contract" Critiques? Medium. https://medium.com/cryptolawreview/whys-szabo-afraid-of-smart-contract-critiques-669ef9e63fc0. Accessed 25 September 2020

Clifford Chance and European Bank for Reconstruction and Development (2018) Smart Contracts: Legal Framework and Proposed Guidelines for Lawmakers. https://talkingtech.cliffordchance.com/en/emerging-technologies/smart-contracts/smart-contracts--legal-framework-and-proposed-guidelines-for-law.html. Accessed 25 September 2020

De Caria R (2019) The Legal Meaning of Smart Contracts. European Review of Private Law 6:731–752

De Filippi PD, Hassan S (2016) Blockchain Technology as a Regulatory Technology: From Code is Law to Law is Code. First Monday 21(12). https://doi.org/10.5210/fm.v21i12.7113

European Parliament (2020) Draft report with recommendations to the commission on a digital services act: adapting commercial and civil law rules for commercial entities operating online. Committee on Legal Affairs, 2020/2019(INL)

Fairfield J (2014) Smart Contracts, Bitcoin Bots, and Consumer Protection. Washington & Lee Law Review Online 71(36)

Felten E (2017) Smart Contracts: Neither Smart nor Contracts? Freedom to Tinker. https://freedom-to-tinker.com/2017/02/20/smart-contracts-neither-smart-notcontracts. Accessed 25 September 2020

Finck M (2019) Blockchain regulation and governance in Europe. Cambridge University Press, Cambridge

Grenon S (2019) Codifying code? Evaluating US smart contract legislation. International Bar Association

Grimmelmann J (2019) All Smart Contracts Are Ambiguous. Journal of Law and Innovation 2(1)

International Swaps and Derivatives Association (2019) ISDA Legal Guidelines for Smart Derivatives Contracts: Introduction. https://www.isda.org/a/MhgME/Legal-Guidelines-for-Smart-Derivatives-Contracts-Introduction.pdf. Accessed 25 September 2020

Jaccard G (2017) Smart Contracts and the Role of Law. JusLetter IT 23

Kaal WA, Calcaterra C (2017) Crypto transaction dispute resolution. University of St. Thomas (Minnesota) Legal Studies Research Paper No. 17–12. https://doi.org/10.2139/ssrn.2992962

Karlstrøm H (2014) Do Libertarians Dream of Electric Coins? The Material Embeddedness of Bitcoin. Distinktion: Journal of Social Theory 15(1). https://doi.org/10.1080/1600910X.2013.870083

Kim AD, Boring P (2018) State-by-State Smart Contract Laws? If it Ain't Broke, Don't Fix It. Coindesk. https://www.coindesk.com/state-state-smart-contract-laws-aint-broke-dont-fix. Accessed 25 September 2020

Kolber AJ (2018) Not-So-Smart Blockchain Contracts and Artificial Responsibility. Stanford Technology Law Review 21(198):214-26

Lauslahti K, Mattila J, Seppala T (2017) Smart Contracts – How Will Blockchain Technology Affect Contractual Practices? ETLA Reports, No. 68. https://doi.org/10.2139/ssrn.3154043

Lessig L (2000) Code is Law: On Liberty in Cyberspace. Harvard Magazine January/February https://harvardmagazine.com/2000/01/code-is-law-html. Accessed on 4 September 2020

Lessig L (2006) Code. Version 2.0. Basic Books, New York

Levi SD, Lipton A (2018) An introduction to smart contracts and their potential and inherent limitations. Harvard Law School Forum on Corporate Governance. https://corpgov.law.harvard.edu/. Accessed 25 September 2020

Levine M (2016) Blockchain Company's Smart Contracts Were Dumb. What matters more: What the code said or what people thought it said? Bloomberg. https://www.bloomberg.com/opinion/articles/2016-06-17/blockchain-company-s-smart-contracts-were-dumb. Accessed 25 September 2020

Long C (2019) What Do Wyoming's 13 New Blockchain Laws Mean? Forbes. https://www.forbes.com/sites/caitlinlong/2019/03/04/what-do-wyomings-new-blockchain-laws-mean/#561b6eda5fde. Accessed 25 September 2020

Low KF, Mik DE (2019) Pause the blockchain legal revolution. Int Compar Law Quart 69(1):135-175 https://doi.org/10.1017/S0020589319000502

Madir J (2018) Smart Contracts: (How) Do They Fit Under Existing Legal Frameworks? https://doi.org/10.2139/ssrn.3301463

Marino B (2016) Unpacking the term 'Smart Contract'. The word 'contract' and Ethereum. Medium. https://medium.com/@ConsenSys/unpacking-the-term-smart-contract-e63238f7db65. Accessed 25 September 2020

Marketsandresearch.biz (2020) Global Smart Contracts Market 2020 by Company, Regions, Type and Application, Forecast to 2025 https://www.marketsandresearch.biz/report/35413/global-smart-contracts-market-2020-by-company-regions-type-and-application-forecast-to-2025. Accessed 25 September 2020

Meyer O (2020) Stopping the Unstoppable - Termination and Unwinding of Smart Contracts. Journal of European Consumer and Market Law 9(1):17 – 24

Mik E (2017) Smart contracts: Terminology, technical limitations and real-world complexity. Law, Innovation and Technology, Research Collection School Of Law 9(2):269–300 https://ink.library.smu.edu.sg/sol_research/2341. Accessed on 24 September 2020

Mik E (2019) Smart Contracts: A Requiem. Journal of Contract Law. https://doi.org/10.2139/ssrn.3499998

Moyle AC, Bacheyeva E (2019) Russian Civil Code Recognizes Digital Rights and Smart Contracts. The new legislation may act as a catalyst for a crypto-evolution within Russian law. Latham & Watkins LLP, Global & Fintech Payment Blogs. https://www.fintechandpayments.com/2019/04/russian-civil-code-recognizes-digital-rights-and-smart-contracts/. Accessed 25 September 2020

Nakamoto S (2008) Bitcoin: A Peer-to-Peer Electronic Cash System https://bitcoin.org/bitcoin.pdf. Accessed 25 September 2020

Neuburger J (2017) Arizona Passes Groundbreaking Blockchain and Smart Contract Law – State Blockchain Laws on the Rise. Proskauer New Media & Technology Law Blog. https://newmedialaw.proskauer.com/2017/04/20/arizona-passes-groundbreaking-blockchain-and-smart-contract-law-state-blockchain-laws-on-the-rise/. Accessed 25 September 2020

O'Hara K (2017) Smart Contracts - Dumb Idea. IEEE Internet Computing 21(2):97-101 doi: https://doi.org/10.1109/MIC.2017.48

Orcutt M States That Are Passing Laws to Govern "Smart Contracts" Have No Idea What They're Doing. MIT Technology Review

Raskin M (2017) The law and legality of smart contracts. Georgetown Law Technology Review 1:305

Reidenberg J (1998) Lex Informatica: The Formulation of Information Policy Rules Through Technology. Texas Law Review 76:553

Reijers W, Coeckelbergh M (2018) The Blockchain as a Narrative Technology: Investigating the Social Ontology and Normative Configurations of Cryptocurrencies. Philosophy & Technology 31:103–130. https://doi.org/10.1007/s13347-016-0239-x

Reijers W, O'Brolcháin F, Haynes P (2016) Governance in Blockchain Technologies & Social Contract Theories. Ledger 1:134–151. http://www.ledgerjournal.org/ojs/index.php/ledger/article/view/62/51. Accessed 25 September 2020

Reyes CL (2017) Conceptualizing Cryptolaw. Nebraska Law Review 96(2):384-445

Rühl G (2020) Smart (Legal) Contracts, or: Which (Contract) Law for Smart Contracts? In: Cappiello B, Carullo G (eds) Blockchain, Law and Governance. Springer. http://ssrn.com/abstract=3552004. Accessed 25 September 2020

Rumi G, Vezzani F, Faelli T (2020) The Financial Technology Law Review, 3rd edn. Italy. The Law Reviews. https://thelawreviews.co.uk/edition/the-financial-technology-law-review-edition-3/1226668/italy. Accessed 25 September 2020

Savelyev A (2016) Contract Law 2.0: «Smart» Contracts As The Beginning Of The End Of Classic Contract Law. National Research University Higher School of Economics. Working Papers Series: Law Wp Brp 71/Law/2016

Sklaroff JM (2018) Smart Contracts and the Cost of Inflexibility. University of Pennsylvania Law Review 166:263

Szabo N (1996) Smart Contracts: Building Blocks for Digital Markets. https://www.fon.hum.uva.nl/rob/Courses/InformationInSpeech/CDROM/Literature/LOTwinterschool2006/szabo.best.vwh.net/smart_contracts_2.html. Accessed 25 September 2020

Szabo N (1997) Smart Contracts: Formalizing and Securing Relationships on Public Networks. First Monday 2(9)

Tashakor M (2018) The New Kid On The Blockchain: Legislative Acceptance of Smart Contracts. Georgetown Law Technology Review https://georgetownlawtechreview.org/the-new-kid-on-the-blockchain-legislative-acceptance-of-smart-contracts/GLTR-04-2018/GE. Accessed 25 September 2020

Tjong Tjin Tai E (2018) Force Majeure and Excuses in Smart Contracts. European Review of Private Law 6:787-904

Tönnissen S, Teuteberg F (2018) Towards a Taxonomy for Smart Contracts. Twenty-Sixth European Conference on Information Systems (ECIS2018), Portsmouth, UK. Research Papers 12

UK Jurisdiction Taskforce (2019) Legal statement on crypto-assets and smart contracts. https://technation.io/about-us/lawtech-panel. Accessed 25 September 2020

US Commodity Futures Trading Commission (2018) Primer on Smart Contracts. https://www.cftc.gov/sites/default/files/2018-1/LabCFTC_PrimerSmartContracts112718.pdf. Accessed 25 September 2020

Verstraete M (2019) The Stakes of Smart Contracts. Loyola University Chicago Law Journal 50(3):743-795

Werbach K, Cornell N (2017) Contracts Ex Machina. Duke Law Journal 67:313-382

Woebbeking K (2019) The Impact of Smart Contracts on Traditional Concepts of Contract Law. JIPITEC 10:106-113

Wood DD (2014) Ethereum: A Secure Decentralized Generalized Transaction Ledger. https://gavwood.com/paper.pdf. Accessed 25 September 2020

Wright A, De Filippi P (2015) Decentralized Blockchain Technology and the Rise of Lex Cryptographia. https://doi.org/10.2139/ssrn.2580664

Milton Keynes UK
Ingram Content Group UK Ltd.
UKHW021106010924
447547UK00004B/6